W0079084

The Executive's How-To Guide to Automation

George E. Danner

The Executive's How-To Guide to Automation

Mastering AI and Algorithm-Driven Business

George E. Danner
Business Laboratory LLC
The Woodlands, TX, USA

ISBN 978-3-319-99788-9   ISBN 978-3-319-99789-6 (eBook)
https://doi.org/10.1007/978-3-319-99789-6

Library of Congress Control Number: 2018958731

© The Editor(s) (if applicable) and The Author(s) 2019
This work is subject to copyright. All rights are solely and exclusively licensed by the Publisher, whether the whole or part of the material is concerned, specifically the rights of translation, reprinting, reuse of illustrations, recitation, broadcasting, reproduction on microfilms or in any other physical way, and transmission or information storage and retrieval, electronic adaptation, computer software, or by similar or dissimilar methodology now known or hereafter developed.
The use of general descriptive names, registered names, trademarks, service marks, etc. in this publication does not imply, even in the absence of a specific statement, that such names are exempt from the relevant protective laws and regulations and therefore free for general use.
The publisher, the authors and the editors are safe to assume that the advice and information in this book are believed to be true and accurate at the date of publication. Neither the publisher nor the authors or the editors give a warranty, express or implied, with respect to the material contained herein or for any errors or omissions that may have been made. The publisher remains neutral with regard to jurisdictional claims in published maps and institutional affiliations.

Cover design by Thomas Howey

This Palgrave Macmillan imprint is published by the registered company Springer Nature Switzerland AG
The registered company address is: Gewerbestrasse 11, 6330 Cham, Switzerland

Preface

It is very hard to pin down the precise genesis for this book. But I will say that it is the culmination of a string of "aha" moments where I sat stunned at my terraced realization that algorithms are the stuff of profitable commerce.

I was formally trained as a mechanical engineer given my childhood fascination with household contraptions and growing up in a gritty, industrial town on the Texas Gulf coast. Soon I found myself collaborating on the design of valves and actuating devices, then broadening my experience with the automation of factories using specialized software during the manufacturing systems craze of the 80's and 90's. Later in life, I attended the Sloan School at MIT, where, without even realizing it at the time, my thinking was changing from a production orientation to business strategy, while weaving in the idea that systems science was equally useful in both domains. This is what I understood well—systems, and I believe that the power of systems thinking is extraordinarily handy in these times of ever-increasing complexity.

My first book, *Profit from Science*, was devoted to the idea and practice of analytical problem solving using the Scientific Method. Research on that book took me around the world to study countless organizations and their approaches to complex business problems from supply chain optimization to smart pricing to competitive strategy to the management of risk. This experience brought me home to the idea that building a company of any size in an enduring way is desperately hard work, littered with lethal traps, and therefore no match for human intuition alone. It takes more than an assembly of smart, capable people doing good work—rather, a key ingredient is the collective intelligence of humans intricately augmented by machines to craft a sequence of competitive advantages year on year. The best firms I observed not only recognized this, but have built deliberate structures, like computational sandboxes, to instill a factory-like institutional capability behind it.

Stephen Wolfram's *A New Kind of Science*, released in 2003, was an astounding work, the result of a decade of hermit-like existence to produce its 800 pages (and half again with end notes). That book taught us all that small programs need not be complex themselves to produce complex, and often astonishingly brilliant results. This fundamentally changed my thinking about who could participate in algorithm design. If small programs could be so powerful, why couldn't ordinary people—those without extensive backgrounds in programming—build small programs? The fact of the matter is that they can, and they should. Stephen Wolfram showed me that remarkable possibility.

As a parent of three amazing children, I felt as though I had a front row seat on the process of learning. Each one unique in his and her own right, bounded through childhood absorbing life in ways that astounded and surprised me. My oldest daughter learning a foreign language, my son learning how to install car audio systems, my youngest daughter learning to ride a horse—all of this coming to them at a remarkably young age. Therein lay even more clues as to how all humans systematically acquire "rules" and effortlessly store them away for later recall.

I think back to the innumerable conversations with colleagues around the world on the grand trifecta of intelligence, technology, and strategy. Stephen Wolfram was himself one of the remarkable thinkers who gave me keen insight into scientific thinking, which overlaid with discussions of strategy with leading CEOs of our time. The juxtaposition of science and strategy led me to the discovery that science and business can energize each other in powerful ways to produce extraordinary value.

Lest you think this book represents a linear path to a single viewpoint, let me correct that for the record—many, many colleagues disagreed with my original thinking, challenged most of my beliefs, and generally told me I had failed to grasp the many subtleties embedded in the simple premise that algorithms could be created by anyone, and monetized by any organization. Thankfully, the thoughts in these pages have been vetted in the crucible of hearty debate among some of the wisest and most experienced business leaders I know.

One of those colleagues, a dear friend of mine from London, Peter Franklin, introduced me to *The Rise of the Robots* just at the time I was thinking about writing this, my second book. To be honest I didn't think much of the premise, nicely summarized in the subtitle: *Technology and the Threat of a Jobless Future*. I consider myself neither influential enough nor wise enough to stave off something as formidable as global mass unemployment. Rather, I read the book with an entirely different goal: how did the author Martin Ford, a successful Silicon Valley entrepreneur, describe a plausible future where the automation of tasks considered "unautomatable" like writing a symphony and grading essay questions become reality? Not just a reality but a dominant feature of the global economy. This really started my mind's gears grinding. I asked myself these questions: If automation is going to be so vital, so inevitable, what will the winners in such an era look like? What can traditional firms do today to thrive in the transformation? Are there entirely new roles and skills required by this future?

The thought experiment led me to conclude that ubiquitous automation is indeed an unavoidable, permanent fixture of the coming economy. If that is the case, then what are the molecules of automation, the ingredients that give rise to it, the tangible objects that are precisely engineered to make the work happen? Clearly, this led me to look at *algorithms* carefully.

The first few Google searches on the subject showed me numerous references to arcane papers on algorithms, all geared to engineers and computer scientists. There was not a single reference to algorithm design and discussion at a layman's level. Not one. Yet I knew instinctively that algorithms were not just for computer scientists—I had spent my whole career methodically extracting algorithms (although they didn't call them that) from human Subject Matter Experts in industries from pharmaceuticals to telecommunications to manufacturing to banking, as a means to create simulation models for those firms. I knew that algorithms and their design could be broadened to encompass ordinary (albeit fearless and forward-thinking) business people. The thing that they lacked was a good set of guidelines and principles. I endeavored to create just that.

And so this book came together, moving from cloudy idea to written manuscript in a fitful, circuitous path. The topic is timely, the lessons vital. Enjoy your journey. Tell me your story.

The Woodlands, TX, USA   George E. Danner
September 2018

Acknowledgements

Being true to oneself means acknowledging the many contributions of others who make any complex work of art possible. This book is no exception to the idea that challenging works are products of an active network. I have had the great fortune to come to know a legion of exceptional people, who possess brilliant ideas and are generous in sharing them with me. Those ideas are the ingredients that this humble servant used to put these pages together in a story that we all hope advances the thinking on an important topic.

The staff at my company Business Laboratory yielded to incessant calls for "hey what do you all think about this" at all hours of the day and night. Alan Savoy, Tripp Lybrand, Miles Hill, and Brooks Lybrand played a pivotal role in my thinking. Our summer interns in 2018, Doug Ziman and Andrew Morrison made meaningful personal contributions to the Digital Twin concept that shaped the core of Part I.

My dear friends from the United Kingdom, Peter Franklin and David Peregrine-Jones are thought leaders with lengthy careers in industry. Simply being in their presence gave me many of the inspirations that guided the theme of the book toward automation. Gerald Ashley set me on this path 5 years ago now by kindly introducing me to his publisher in a fateful move.

Mark Serensey, a Manager with Oil States Energy Services is probably one of the most well-read individuals I have ever met, a polymath by any standard. Many of the references used here came from Mark's keen eye cast over the vast universe of written media in the field.

I hereby dedicate this book to Mr. Andrew Palmer, CEO of Idarat Reslience. Andrew and I became fast friends through a network of mutual acquaintances, and our long walks on hot Jumeirah Beach in Dubai near his home guided my thinking as the book was starting to unfold. Sadly, Andrew passed away of cancer in August 2018, just as the writing was coming to an end. Andrew, I hope you are proud of what we've done. Your fingerprints are all over it. Rest in peace my good friend.

This book came about during a tumultuous time in our family life, marred by divorce. My children Starr, Grayson, and Sage unflinchingly remained supportive of my dream to write about topics that are close to my heart and reflective of my life's work in spite of our family's challenges. My little band of happy warriors kept me smiling through that dreadful process. Kids, I love you all with every fiber of my being.

I am grateful to God that He gave me the gift of writing, then put people in my life that helped me make a small difference in the world. Lord, I humbly ask your continued blessings on us all.

Contents

List of Figures

Part I

Our Automated Future

This morning I boarded my flight from London to Houston. An hour before I checked my bag, and received my boarding pass. I walked onto the plane, took my seat. I am barely aware of the engines as they hum away at 35,000 feet. I switch on a movie, and settle in for the 9 hour journey.

So it seems that today was just like countless other ordinary days in my life...or was it? In fact, at each and every step there were millions of calculations going on around me, under my notice, making the morning's events flow effortlessly one to another. It is a software-driven world to be sure, but software and information technology are just the delivery mechanisms. What *actually* made all of these sophisticated logistics, machine, and security functions happen?

In a word: rules.

A complex tapestry of rules encoded into the airline's reservation system delivered my boarding pass to me. The system recognized me as a frequent flier, and applied a whole different branch of rules to my reservation based on that status. I have no doubt that a dozen security cameras locked onto me as I walked through Heathrow, churning through a book of rules to determine my intentions. Once on-board, the flight attendants followed a well scripted set of rules from seating passengers to security announcements to closing the doors. The flight controller's rules took my weight into account and told the engine to provide just the right thrust to keep me and my fellow passengers aloft. Unseen, a sea of computations following prescribed rules that would fill hundreds of libraries worth of written pages surrounded me, enabling every mundane facet of my movements this morning.

How extraordinary it would be to have "Z-Ray" vision sufficient to *see* all of that code in real time coming together, as it moved me along!

Martin Ford in his book *The Rise of the Robots* paints a dramatic picture of the future—one characterized by pervasive automation, including automation of creative and inherently human tasks like writing a novel or passing judgment in a legal case. Many other influential authors and thinkers have supported this view.

When an entire set of rules are encoded in technology, we refer to them as *algorithms*. Algorithms are not new, and in fact date back to ancient times. Ciphers, simple algorithms to substitute alphabet characters used in primitive cryptography so that armies could pass secret messages to one another were discussed in the 4th century BCE.

Therefore, if the hallmark of the future is automation, and algorithms form the "molecules" of an automated system, one cannot escape the inevitable conclusion that those who participate in the design of algorithms will thrive in the coming economy. The new idea that I put forward in these pages is that algorithms are no longer the sole province of computer scientists and programmers. Ordinary people, most especially ordinary people involved in running a business or an organization of any stripe, can and should participate in the design of algorithms.

Algorithm design in the way that I will describe in this book will be one of the leading business skills of the coming decade, easily surpassing today's most popular analytical skill known as data science. I will begin by providing a historical context for algorithms, looking at "famous" algorithms across history, some created centuries before the invention of the computer. From there we will move on to understand the precise role algorithms will play in a highly automated commercial future, enabling smart business models and overturning certain industries. I will invite you to try your own hand at algorithm design, describing the means by which everyday people, not programmers (although programmers will enjoy this as well) can build, test, and validate their own algorithms. Since algorithms on paper may be beautiful to behold but don't do anything useful, we will address the important work of implementing algorithms within a technological system. Many knowledge workers in a company coming together to create algorithms to drive efficient operations or enhance new products suggest an almost factory-like nature to algorithm design for institutions. The factory metaphor is no accident—we must rethink the way corporations are stitched together to encourage greater capacity for algorithm design, notwithstanding their protection, archival, and monetization. Finally, I will offer some lessons as to where this newfound intelligence will take us, and what it means to stay competitive in an environment of ever-increasing technological sophistication.

I am delighted that you've chosen to join me on these pages. I will reward you with a number of surprises—things you never knew about your own potential to participate in a lightning-fast economy. I promise to leave you with a set of enduring skills that you can apply right away. But most of all I want to pay forward the gift my mentors in life have given me—the gift of keen insight, a way of thinking that you did not possess before; leading us to a new optimism about our collective future in a world where we will harness intelligence for good, yielding benefits beyond our imagination.

Welcome, friends, to the dawning of a new era in computation.

Let's begin.

1

Automation Is Here

The first century A.D. was an exhilarating time in Alexandria, Egypt. The happy collision of engineering and mathematics of ancient Greece with the carefully curated knowledge of Egypt gave rise to an amazing number of sophisticated machines, even by today's standards. It was at this time that a prominent engineer and mathematician, Heron of Alexandria, created the very first vending machine.

Much of the income derived by the priests of the many temples around Alexandria came from the sale of religious elements to the temple-goers, such as parcels of holy water used to wash the face and hands. Priests were notably dismayed by certain visitors taking more water than they had paid for at the temple entrance. Heron devised a simple but ingenious device that allowed a coin to be inserted into a vessel. The coin fell upon a pan that was at one end of a balanced beam. The weight of the coin raised the opposite side which was attached to a valve that opened to allow the water to flow out of a tube at the bottom of the vessel. When the weight of the coin equalized with the weight of the water, the valve subsequently closed, giving the temple-goer a precise amount of water in accordance with what was paid [1].

Eighteen hundred years later (!) Percival Everett invented the first modern age coin-operated vending machine to dispense postcards, primarily at railway stations across London. Over the next two decades that followed, the products in the machines expanded to include gum, cigarettes, chocolate, and soap. Small, incremental improvements were made over the next seventy years until the advent of the microprocessor and communications networks, thus allowing vending machines to accept a much wider variety of payment

© The Author(s) 2019
G. E. Danner, *The Executive's How-To Guide to Automation*,
https://doi.org/10.1007/978-3-319-99789-6_1

forms outside of physical coins, as well as serving in two directions, both vending and receiving, as is the case with library books and DVD rentals.

At the one and only Amazon Go retail store in Seattle (as of this writing), a customer enters, picks up items, and leaves the store. The "checkout" is fully automated using an army of unseen cameras and sensors, all meticulously coordinated under a concept known as *sensor fusion*. Amazon plans to roll out many more stores after learning from the initial opening [2]. Heron would be quite impressed.

A fascinating contrast of history with the modern age… but what does it mean? In all cases, we have a customer that wants something, and is willing to pay for it. We have suppliers willing to provide the items at an agreed price. The stories I've told here revolve around the mechanism for making the transaction which is designed for one purpose: to minimize the *friction* of the transaction. Friction can come in many forms—theft in the temple, availability and cost of store clerks to sell gum, and long lines at grocery stores. Clever machinery (technology) gradually removed friction, solving the customer's problem, whether they were conscious of a "problem" or not. Transactions now happen faster, cheaper, more accurately, and with greater optionality than before.

This is automation.

As you can see, it is not a new concept. What is new is our limitless ability to infuse automation into business models, not exclusively by an elite group of technologists, but by anyone with a *systems mindset*.

Two decades of technological advance in four specific vectors has placed us at a unique inflection point in the history of automation. First, sensors for everything from temperature to motion to image recognition are now lower in power consumption, cheaper to buy, and of a resolution and accuracy that allows computers to know in astonishing detail many aspects of the physical world around it. Computing has broken free from the bounds of "computers" to run almost anywhere, and at a speed sufficient to do very sophisticated calculations, such as processing an image to find and identify human faces. Data is now universal, accurate, and ubiquitous, where once we had precious little. Moreover, we have well-proven sharing mechanisms that we can use to voluntarily distribute our specific data intentionally to trusted partners.

But perhaps the most important development giving rise to the automation boom is in the sophistication of the *science* of automation. In other words, we can harness computing to do the work that humans do, by mimicking the same processes that the brain uses to make decisions. Artificial intelligence is the cornerstone of this new, more comprehensive style of automation. There is no question that AI has enjoyed a renaissance in the 2010 decade, alongside a greater awareness (and less fear) of how the technology can be deployed to practical use.

History has taught us that real inflection points in technological advances come along when a suite of seemingly unrelated threads blend together in creative and ingenious ways [3]. When the printing press paralleled advances in naval architecture and design in Renaissance Europe, books could not only be printed but they could be put on vessels to carry that knowledge from one country to the next, unleashing an unprecedented wave of knowledge and invention. Today sensors, computing, data, and AI are in forms that are low cost and practical. It just so happens that these are the key ingredients in weaving automation into businesses to make them perform better, faster, and cheaper. Over the next decade we will witness an equally potent wave of automation, overturning commonly held beliefs of "oh we will *never* automate that" again and again.

Even today we have intelligent systems *autonomously* dispensing tax advice, diagnosing medical conditions, and optimizing farm yields where once we had humans holding forth with the aid of data and computers. Automation is upon us, but in hidden, out-of-the-way corners of the economy. The full wave is yet to start, but I see the 2020 decade as the staging ground for it. When it comes, intelligent systems will be the norm, the default expectation, rather than the amusing exception that exists today.

The decades of the 1970s and 1980s were filled with examples of *physical* automation. The automobile manufacturing industry dove head first into the concept of the "lights out" factory with robots employed across the assembly line. Semiconductors and consumer goods followed suit. The coming wave will be distinctly different—not focused on physical work, but rather on "white collar" processes and functions that involve judgment and reasoning on knowingly incomplete data. In doing so we can set aside the complicated science of kinetics and kinematics needed in the physical world of work and focus exclusively on logic and data as a means for computers to make important decisions. Leading thinkers in business science have coalesced around a set of ideas collectively referred to as *Industry 4.0*, suggesting that we are in a 4th wave of commercial power and enlightenment since the Industrial Revolution of the late 1700s.

The building block of new automation is the algorithm, a clear and precise description of how a decision process works, not expressed in code but rather in words and pictures that are intended for a broad audience of collaborators. This is perhaps the fundamental principle of this book—algorithm design is not by any means the sole province of computing specialists but is accessible to everyone. In these pages we will show you how to design, build, and test a special class of algorithms that are used to automate systems (mostly business functions). In fact, this book will make thin distinction between automation and the algorithms that underlie automated systems.

Our global economy will need vast armies of automation practitioners, at many levels, just to keep the industry moving along. Where will these talents come from?

One answer to the question lies in the present focus on analytics and data. The landmark book *Moneyball* by Michael Lewis showed us that even in the tradition-bound industry of professional sports, analytics is a potent competitive weapon. This in turn ushered in an era of analytics to the extent that a brand-new specialist role emerged in corporate settings called *data scientist*. These folks were released on an organization's data and processes and were expected to lead a digital transformation of the company from undisciplined and manual to data-driven and algorithmic. In many cases this worked rather well for the sponsoring organization.[1]

As we write this book in 2018 data and analytics is enjoying a wave of popularity and interest. Every company in the world expresses the same ambition: to be more data-driven, to be more disciplined it is decision processes, to codify the tacit knowledge that exists in human brains alone. The very best companies are working on this in earnest, in a formal, visible way. That leaves a large number of companies sitting outside looking in desperately wanting the benefits that analytics brings. The guidance expressed in this book will take companies on a journey from outside to inside along a very practical path.

Building an analytical model to solve some complex business problem involves a process that starts with ahypothesis—a concise restatement of the business problem, then moves on to the creation of a series of diagrams collectively referred to as a qualitative model. The qualitative model serves as a developer's blueprint for the quantitative model, which additionally uses data to make the model "come alive" analytically. Finally, analysis is the controlled experimentation with the model to explore a wide range of future scenarios [4].

What are the steps in automating a complex business function, such as a supply chain or in pricing a product? This involves creating a concise finan-

---

[1]There are indeed a few cases where he data scientist label was misapplied to people who have no business being data scientists in the true sense of that phrase.

cial/operational motivation statement regarding the nature of the automation proposed, then carefully diagramming the way this function works in current form, building the associated algorithms, and testing the resulting system under a wide variety of conditions. In other words, very similar to the same process used in analytics!

It is fair to conclude that many of the people who are skilled practitioners of analytics today will likely be repurposed to become automation builders, people I will later refer to as practitioners. Demand will first be satisfied along this path of least resistance. Repurposed talent will be the initial source of talent to prime the pump, but we will need many, many more souls to bring to bear on automation in the general, global economy. Students in school today would be well advised to learn the complement of skills needed to be successful participants in the automation wave. Students everywhere must develop a *systems mindset.*

The flip side to the human equation in automation is mass unemployment. Today it is fashionable to write tales of automation as an evil force, rendering human employees without vocations giving them a life sentence of poverty and desperation. Fashionable, yes, but not grounded in truth.

In my work I get a chance to meet with many, many senior leaders of companies large and small. Mass firing of humans through automation has not come up in a single conversation I've had in the last decade. Rather, the vast majority of asset operators wish to take the existing human workforce and have it produce 3×, 5×, or 10× what it had produced at the baseline. This philosophy makes the value of the human workforce in place *more* important, not less! It also means that hordes of humans huddling in cubicles typing numbers into Excel spreadsheets will be a vestige of the past. Human workers do an astonishing number of mindless tasks every day.

As a reader of this book, you have two choices. The first is to amuse yourself with our stories, but take no tangible action until the evidence is so astoundingly clear that action must be taken as a defensive step. Yellow Cab took this approach, dismissing Uber and Lyft as minor league niche players that would never amount to a competitive threat. By the time Yellow Cab understood the implications, a coveted taxi medallion in New York City, once a golden ticket, had collapsed in value [5]. The speed of automation advances parallels the speed of technology-enabled business model innovation. Taking no action comes at considerable risk.

The second choice is to act, <u>now</u>. Even though the bigger wave of automation is still a few years away, there are health-and-hygiene steps that can be taken now as an easy on-ramp to tomorrow.

1. Build an ideation engine to foster clever thinking about ways to automate the business
2. Master the art of understanding systems (any system) through diagramming
3. Create sleeper cells of automation talent that can spring into action quickly as the wave approaches
4. Develop a working knowledge of the technology tapestry of automation
5. Make a science of studying other firms in industries completely different from your own
6. Compile good quality data, but do not allow the data collection effort to dominate the work
7. Build a sandbox for experimentation with automation concepts

These steps align the chess pieces inside an organization in a way that makes the inevitable scale-up later on much more efficient and less disruptive.

Who will do this work? My assumption here is that you as the reader of this book are deeply interested in the subject, so much so that you want to take action for the benefit of your firm. If you are one of those people I have a name for you: **practitioner**. These are people who, irrespective of their job title, are motivated to learn the principles of automation and apply them rigorously to create a Version 2.0 of the firm as it sits today. Practitioners do not have to be programmers, nor technologists. Rather they simply need to have a systems mindset to effect automation in their companies. This book was written for you practitioners out there struggling to make a difference for yourself, your industry, and your firm. Subsequently this book is also written for senior leaders who might be responsible for making step changes in the organizations performance by hiring practitioners into a team, giving them a sandbox to play in, and directing them toward useful work.

So what is a systems mindset exactly? How do I know if I have one already? If I don't have one how do I get one?

A systems mindset means you naturally, instinctively think in systems. Your brain is wired in such a way that when you move through the world you organize things that perform work in various systems that in many cases are well integrated with another. Say you read an article about global trade. Companies in a given country are systems that operate according to rules that are generated by sovereign governments and consumers then produce the derivative products from these companies and so on. Perhaps the article references how the trade rules are changing which in turn affects the machinery of global trade for better or worse. A person with a systems mindset effortlessly converts the text of the article into a network diagram of the actors and their relationships in the same way that a subsystem diagram of a car very clearly shows the relationship between the engine and the braking system and the electronics and the transmission. More formally, this is the work of a whole body of science called *Systems Thinking*, pioneered at MIT in the early 1950s. Systems Thinking is the fundamental framework that underlies all models of real-world systems. As such it is the catalyst for effective automation, as automation seeks to "model" the action of a human performing the same task.

Even passive observers of the economy recognize that the evidence that automation is here on our doorstep is undeniable. But that doesn't tautologically imply that the degree of automation will continue at an accelerated pace, as I am suggesting. Rather, automation will be propelled forward by a simple imperative: economics. Unpacking the economics of automation will give us insight into how it should be harnessed for sustainable business value. The financial incentive to automate is a powerful, incontrovertible force. There is nothing we can do to stop it even if we wanted to. The only thing we can do as a society is to determine how best to thrive in the coming automation era. This is what this book is about.

Summary

We sit on a unique moment in history. The past has brought us technological innovations that allow us to very closely mimic the complex decision-making process of humans. It has also produced organizations with vast numbers of complex problems to solve every hour of every day. Up to now most organizations operate with an army of humans somehow making it all work by the end of the day.

The future will be different. Very different.

Because we possess the potential for automation that we did not have before, we are set to embark on an unprecedented wave of automation for organizations of every shape and size. This book is purpose-built for getting you ready for that highly automated future. To do that let us first examine the economic imperative of automation, because in doing so we will better understand the shape of automation that is most likely to gain traction for our organizations.

Bibliography

1. https://en.wikipedia.org/wiki/Vending_machine.
2. "Amazon Go Means Goodbye Status Quo", *Forbes*, July 6, 2018.
3. Walter Isaacson, *The Innovators* (New York: Simon & Schuster, 2015).
4. George E. Danner, *Profit from Science* (New York: Palgrave, 2015).
5. "Taxi Medallions, Once a Safe Investment, Now Drag Owners Into Debt", *New York Times*, September 10, 2017.

2

The Economic Case for Automation

The very word "automation" these days conjures up notions of human lookalike robot armies doing jobs in the same way that humans do them, displacing those very workers who are left aimlessly walking the streets looking for meaning. While that may make a popular movie or TV show, it belies the reality of what automation will do in the coming decade. True, human roles will shift. But history teaches us that economies always find a way in the long run to absorb radical new technologies, and there is no evidence that I've found to suggest that automation will forge a different path from the steam engine or the personal computer or the printing press.

Let's start with the domain of automation—physical versus cognitive. A dramatic rise in physical automation occurred in the late 1970s through the mid-1980s with the US and UK automotive industry's response to Japanese production, most notably from Toyota, a bellwether of automation. The Japanese not only created highly automated factories but also matched this brilliantly with efficient methods of work to create a remarkable advantage in a global industry. This one-two punch from Japan proved highly effective and Western automotive firms sought to emulate that by building factories governed by philosophies of "lights out", Just In Time, and cellular manufacturing, later folded into a set of concepts under the umbrella of lean manufacturing. Gradually the labor content per car declined. Where once humans welded pieces of assembly and painted exteriors, now machines did almost all of the work. Humans simply kept the machines running. Today this level of automation is the norm, and few observers bat an eye when they see robotic movements of materials through any common factory.

© The Author(s) 2019
G. E. Danner, *The Executive's How-To Guide to Automation*,
https://doi.org/10.1007/978-3-319-99789-6_2

As online retail has grown, physical automation has found its way into the management of warehouses and distribution centers—the extent of which is unparalleled. It isn't a stretch of the imagination to think about Amazon-like fulfillment centers operating without a single human present in the near future.

It is surprising that physical automation was the first generation of automation to come into view. Physical automation requires one to overcome the kinematics of balanced movement, which any engineer will tell you is a very hard problem to solve. And the payback once you've solved that problem? Physical automation displaces blue collar labor, a significant but lower cost strata than other labor categories that exist in organizations. While an interesting piece of our economic history, physical automation is a far less important topic today than the other side of the coin—*cognitive* automation.

Do you frequently walk the halls of organizations? I do, and it is a fascinating experiment. What I find is that the vast majority of actual work that human professionals are doing is filling out spreadsheets, typing into web forms, writing emails to relay some piece of information, engaging in a workflow (like approving some expense), and so on. In other words, humans today are engaged in astonishingly mechanical, repetitive *electronic* processes that even simple algorithms could perform. And this so-called professional work is the highest strata of our labor costs. What a wasteful use of our limited human talent.

In later chapters of this book we will discuss in detail how one goes about taking these white-collar processes from manual to automated. For the moment let us focus on the economic changes that ensue from cognitive automation.

Jobs will morph from "doing a job" to "thinking about doing a job". In other words, people will be free to think creatively about the larger, end-to-end picture of their role in the company's value creation, and engineer ways for that job to run better, faster, and cheaper. In fact, it is not infeasible for one's job to be to eliminate their own job (and such a person should be richly rewarded for doing just that—we call this a *gainsharing contract*).

When I speak to senior executives all around the world in every conceivable industry, they all tell me the same thing: the motivation for automation is *not* the complete removal of large swaths of human workers from a company. Rather, company leaders want the existing corps of human workers to do $2\times$, $3\times$, $5\times$, or even $10\times$ the production they did before. The emphasis is on the numerator (output) side of the productivity equation. That makes the people who work for you *more* important, not less! I do not buy the thesis that is common these days, even among writers and thinkers who should know better, that the future will be punctuated by mass human unemployment and intractable wealth disparity. Temporary displacements will occur and are unfortunate, but overall our economies will hum along at near-to-full employment for the foreseeable future. Automation and its benefits will have a stabilizing, not a disruptive effect, because the labor force will be a smaller, leaner component of the Profit and Loss statement, and therefore cannot be used a lever for cost cutting through boom-and-bust cycles that we've seen in many industries.

Couple this with the fact that automation is a relatively inexpensive endeavor compared to many other kinds of investments that organizations typically make. In later chapters we will stroll through a vast tapestry of technology that underlies automation, many of which come at a surprisingly low acquisition cost. While it is true that the application of automation requires professional talent that isn't particularly cheap, the denominator in the productivity equation is easily manageable. That has additional implications to the fact that smaller companies can employ automation just as effectively as the large caps, even non-profit organizations.

Return on Net Assets (RONA) has become the telltale sign of smart management for industrialized businesses. It is one of the most reliable aspects of an organization's report card. A good RONA simply says, "I took the same dollars in fixed assets that my peers have and managed them for value better than they did." This applies to both industry peers as well as best-in-class firms that extend across industries. A before-versus-after RONA for the same company (in two or more contiguous time periods) also shows the benefits of an effective automation strategy and execution roadmap. In short, a good RONA is the reward for good automation practices. Particularly for industrial firms, the better ones will make a serious effort to elevate RONA into the top metrics that guide the company.

What about non-industrialized firms or service firms that don't own substantial fixed assets? The benefits of automation to these firms can be substantial, as such firms tend to be more human labor-intensive than their industrial counterparts. There is a clear financial return to reducing operational expenditure by having the same labor pool accomplish n-times-X more work output than before the automation was put into place, but at the same time there are no commonly used financial ratios that isolate the effect of automation alone.

I believe that innovations in the accounting and finance world will catch up to this gap and will eventually devise financial ratios that distinguish the automation capabilities among firms. Economists long ago have recognized the importance of *multi-factor productivity*, most notably the Nobel Prize-winning economist from MIT, Dr. Robert Solow. Dr. Solow devised a simple but ingenious equation called the Solow Residual, which stated:

$$\text{Output} = \text{some function of (Labor} + \text{Capital)}$$

therefore:

A change in output = a function of (a change in labor + a change in capital)

The function—which could be a simple constant or a rather complex thing—is a reflection of how well the firm uses its labor and capital and could be thought of year on year, or company to company as to how adept the organization leverages both labor and capital in intelligent ways.

Dr. Solow's work gave rise to a whole new discipline of accounting known as *Growth Accounting*, which emphasizes the importance of investment in machines and technology to secure future financial returns.

My belief is that as a community, the accounting profession will pay greater attention to the revamp of accounting ratios to more accurately reflect the winners and losers in the automation race of the coming years. I could forsee Solow-like thinking giving rise to a thoughtful, objective measure like *Return on Human Capital* or *Return on Knowledge Capital*.

Taxes on automation could slow (but never stop) its forward progress. The perplexing idea of a "robot tax" has been floated by notable people as a means of financing a guaranteed basic income to those displaced by automation. Not only is this a bad idea, the implementation challenges are enormous—is a simple algorithm a robot? Where does it start and stop? While I don't see this ever amounting to a serious challenge, it is something for the financial community to watch as it unfolds.

Now let us look at all of this through the lens of a tangible example. Today we have a front row seat on a classic economic case for cognitive automation: self-driving trucks.

The United States has a big problem: there are not enough truck drivers here to satisfy demand. And it's getting worse: truck driving schools are not turning out graduates in sufficient numbers to meet the coming demand, much less the existing gap. This is a subtle but hidden brake on the growth of our economy. Who knew?

Driving a truck is a lonely, tedious profession. I have no doubt that there are many truck drivers out there in the world who love their jobs—however it is a fact that the younger generation just coming into the workforce is not embracing truck driving because of its long hours and difficult working conditions. Relying on more human drivers alone will not fulfill the demand.

Enter the self-driving truck.

A host of companies today are aggressively pursuing the notion of a commercial payload-bearing truck that drives itself in some manner. Companies like Embark, Volvo, Daimler, nVIDIA, Waymo, Peloton, Tesla and Otto are part of a rapidly expanding ecosystem, fueled by keen interest from investors who have funded these firms handsomely. Embark has formed a partnership with Ryder, a pattern that we see as common to this space whereby a legacy trucking firm joins up with a technology provider to form a comprehensive offering to the transportation market.

nVIDIA meanwhile is making the chips that integrate numerous cameras and other sensors at astonishing speeds to perform the complex task of navigating an 18 wheeler. Waymo was spun out of Google's self-driving car division. Uber bought automated truck technology company Otto to complement its new Uber Freight business. The cost of a retrofit of a standard truck to an automated one: around $30,000 USD according to Uber.

In perhaps one of the most fascinating stories of all, US Xpress, a sizable trucking firm, is now partnered with another firm called TransRisk, which collects market data in real time with the intention of serving up relevant data to an automated truck in motion [1]. Max Fuller founded US Xpress over 30 years ago. TransRisk, was founded by Max's 38-year-old son Craig. Says Craig about the relationship, "I'm betting on data and information services. He's betting on owning equipment." Welcome to the way the new generation is thinking.

Self-driving trucks is not a one-and-done process. Like all automation of critical systems, especially those where human safety is involved, will develop in degrees. One of the first models of self-driving trucks is the so-called platooning system. In platooning a lead truck is manned by a human driver. Behind the driver a train of between four and six driverless trucks follows closely, perhaps within a few feet, all of which are coordinated in real time using a wireless network. Platooning not only cuts down on drivers per unit load delivered, but also reduces traffic congestion because of the tight spacing. It also allows the train of follower trucks to ride on the air draft effect from the lead truck, resulting in considerable fuel savings. It has been said that platooning trains could bring trucks competitive dollar-for-dollar with railroads, which would be an incredible feat.

The next model to emerge might also be the Bar Pilot model. Since highway travel is relatively safer and more controllable than local roads, the self-driving truck handles the long haul portion of the journey, while human drivers navigate the truck for the first and last miles of the journey, enabling one human driver to spread across many different deliveries.

While fully ghosted operation of a truck is inevitable, these kinds of Human in the Loop (HIL) models are the stepping stones along the way.

We present the story of the self-driving truck not simply as a fascinating anecdote, but rather as a classic case study in how an economic impetus led to an ecosystem which led to a technology tapestry which precipitated investor favor which is ultimately leading to degrees of automation unfolding on the highways before us. The self-driving truck narrative holds universal lessons for us in how automation will progress. My guess is that your industry, whatever it is, will see the same kind of dynamic. If your governing board is not already engaged in thoughtful conversations around the overlay of automation on the company and its industry, it certainly should start. Now. Right now.

Because the economics of automation are so compelling, it is crucial to make newly initiated automation a *business* project, not a technology project. Technology is an important enabling force for automation, but it should never be the driver. I have personally witnessed the failure of many projects for this very reason. Hand over automation to the IT department and it will be guaranteed to have a technology focus and an obsession with data, rather than the proper emphasis on the business problem that the automation is intending to address.

The proper way to do this is to create a business hypothesis for the automation. A hypothesis—the principal tenet behind the Scientific Method—is a clear and concise restatement of the problem that you are trying to solve. The hypothesis only lightly hints at the solution; the emphasis is on the problem one is trying to solve through the lens of automation. I cannot underemphasize the importance of a good hypothesis in creating a successful automation outcome. The hypothesis becomes the governing document to guide later development and to keep the scope in line with the spirit of the business challenge. It is so easy and tempting to steer away from first principles with the excitement of automation's possibilities and to become enamored with the exotic technology tapestry that is often brought to bear in these projects. Good automation project discipline requires that we *work backwards* from the business problem versus working forwards from the technology.

A good hypothesis has the following characteristics:

1. **It is graphical**. The purpose of the hypothesis is to communicate quickly and efficiently to people well outside the zone of expertise of automation. Therefore, the more we can make the hypothesis a picture, the crisper the communication to our audience.

2. **It is simple**. Our goal is to capture the essence of the automation question, not provide yet more noise to the question with inappropriate detail.

3. **It is agnostic on the solution**. We want to express the business problem addressed by automation, not design or even suggest a design for the ultimate solution—yet to be worked out by the implementors.

4. **It incorporates time**. Almost all complex business problems unfold over time, with short-term effects distinct from long-term ones.

5. **It suggests the extent of the prize**. Assuming this business problem is solved, what kinds of gains could one expect to see?

6. **It incorporates a contrast against the "do nothing" case**. Corporate inertia will likely dictate that "do nothing" is the default decision. How will the future be different if we proceed? Are there risks associated with the "do nothing" case?

A well-written hypothesis anchors the automation to a specific business problem, and the solution to that business problem has an economic return associated with it. Those who seek to apply automation effectively within organizations should develop a habit of building hypotheses as the governing mechanism for automation.

Technologists, on the other hand, play an important role as enablers. In my view the IT team should be the owners of what I call *The Art of the Possible*—a notional list of technologies and what they do well. It can be said that knowing The Art of the Possible may actually shape the kinds of automation that are built, and I fully endorse a thought process that seeks to link elements of The Art of the Possible with the other parallel list of complex business challenges faced by the organization on any given day. Note the subtle difference here between allowing IT to lead automation versus the more correct role of acting as curators of key technologies.

For financial investors of all shapes and sizes, automation presents opportunities—lots of them. But like any investment strategy, you must play the game well to succeed.

Investor's goals are simple: buy assets that have long-term value and sell them when that value has been realized. The trick is in predicting which assets hold the most future value. Starting today, out to the foreseeable future (10 years or so), one of the most successful investment strategies will be with companies that are in traditional, unautomated industries that deliberately embrace automation, thereby changing the entire business model of the firm. Facebook and Google are not good examples of this, as they are pure technology plays, however, Amazon's retail business and Netflix are excellent examples of automation and algorithms playing a pivotal role in their business models as distinct from traditional retail stores and entertainment companies respectively. In a sense, companies like Amazon foretell what steelmaking, chemical production, and energy generation could look like in a decade's time.

If you are part of *any* organization that has investors, you should be aware that the performance bar has now been raised for you. In the past, the concept of sector investing dominated the thinking among the investor class: those who invest in oil and gas, for example, choose to allocate their funds to the best performing oil and gas firms. Same for real estate, transportation, and a host of other definitive sectors. Therefore, as a firm, your job was to outperform your peers. If you were in a slow-moving industry headed by unimaginative, caretaker leadership, outperforming peers was not too difficult.

The sector investing approach has fallen away in recent years. Investors now feel free to choose between company A and company B regardless of industry. What that means is that Coca-Cola competes with Apple for shareholder capital, irrespective of their radically different businesses. This extends even to privately held firms. One of our company's clients is a rather large set of industrial businesses owned by a wealthy family. The new generation of business leaders within the family is openly asking questions about whether they should continue to own these businesses when other investments are offering more satisfying returns at even lower risks. They are not wrong to raise these questions.

Add it all up: companies cannot rest on being just as mediocre as their peer down the street any longer. They have to perform well as measured on a broader scale. While there are several remedies to the performance gap problem presented by this, automation certainly represents the kind of step change that is necessary to allow, say, a mining firm to be just as attractive to an investor as a technology or a media firm.

It is easy for investors to fall into the trap of putting their dollars behind the underlying technology for automation, rather than the exploiters of the technology (the sexy technology gets all the media attention). This is akin to investing in the steam engine in the late 1700s versus an identical investment in a railroad. While steam engine investors would likely be rewarded well enough, history shows us that railroad investors capitalized on steam engine technology to a much greater extent and earned far and away more money.

For institutional investors, the picture is even more intriguing, as this domain allows firms to actively engage in the automation strategy. In the

Private Equity asset class, for example, General Partner (GP) firms buy up a set of companies assembled into a portfolio, and typically hold those companies for six or seven years before divestiture. In that time, GPs will often inject expertise and large cap company-style sophistication into the portfolio company as a means to enhance returns to the firm. This is commonly known in the industry as Operational Alpha, the latter word referring to the supernormal returns that are earned by a deliberate strategy behind the original buyout.

To put this in context, this investment strategy means that the GP would seek out under-automated firms in relatively sleepy, slow-moving industries and add a "just right" surgical dose of automation and algorithmic functionality to transform the company to a much higher level of financial performance. Moreover, the algorithms that do the crucial work of the company are no longer sitting in human brains but in fact are carefully curated and documented as the intellectual property (IP) of the firm. That means that not only does the firm simply perform better than its unautomated self outright, but that performance could be scaled up to an even bigger version of itself. It could even be exported to other dissimilar firms in the portfolio, not to mention that the automation IP will at some point become a first-class citizen on the balance sheet. In fact, I could foresee a strategy where a GP would buy a portfolio company for the sole purpose of learning how to automate it, then use that learning to acquire other companies that can exploit the same style of automation. Large cap firms could use this as well in their M&A activities. This is an aspect of investment playbooks that is not yet widespread today. Stay tuned.

I do not wish to leave you with the impression that the economic picture is all one-sided in favor of automation. There are indeed some headwinds that all automation practitioners must understand and be prepared to manage.

The first is the fact that a substantial portion of the benefits of automation is not well measured by our managerial accounting systems, so the financial value is often hidden. Automation allows us to become more agile in processes, and deal with sudden surges (or valleys) in demand. This capability value is rather hard to measure, and hard-nosed executives with little tolerance for leaps of faith may dismiss an automation proposal that falls short of an expected internal rate of return. The option value of automation is often very high, so my recommendation to its sponsors is to incorporate the Real Options [2] approach into the overall picture of value that is presented to a senior leader.

The second challenge is the accounting and tax treatment of automation. Automation IP is not (yet) formally recognized on the balance sheet, merely falling into an amorphous bucket called goodwill. All accounting experts agree that IP is valuable and should be measured. None have come to an agreement on how to do it. I believe that this problem will ultimately be solved. As we speak there are corners of academia that are dedicated to a solution.

Taxation is a far more concerning issue that we've raised in the previous chapter. Certain politicians have made public mention of taxing automation goods and processes as a means to fund those citizens who would ostensibly be negatively affected by automation. This is a wholly bad idea, not to mention the complexity of enforcing such a law. Yet, it remains a conversation in humanity's circles and will not likely go away, given politician's bent to playing to the worst of our fears of a *Terminator*-dominated society. On the plus side, I believe that in the automation era we will see less race-to-the-bottom tax incentives being offered to companies to relocate to a given place. I for one am stunned that municipalities are being suckered into this kind of one-sided game, but the headlines are littered with countless stories on any given day.

There is no doubt these are non-trivial challenges. I do not believe the challenges will halt or even slow down the inevitable march of automation. However, for practitioners the wise approach is to be aware that these roadblocks exist so that you have the intellectual muscle to step over them as they arise.

Summary

The economics of automation are more than compelling—they are inevitable. There is nothing that you nor I can do to stop the forward march of automation to unprecedented levels. The things we thought could never be automated, will be automated via the business imperatives that we've just discussed. The only choice you have to make here is the extent to which you are an active participant in this future highly automated world.

But what is automation, really? What makes it up? How does an organization "do" automation, given that you've bought into my argument about the economic benefits? The answer lies in the fundamental unit of work of automation: fractional pieces of software code called *algorithms*. We've mentioned algorithms several times so far, now let's see how they really work within the automation lexicon.

Bibliography

1. "Self-Driving Trucks May Be Closer Than They Appear", *New York Times*, November 13, 2017.
2. Martha Amram and Nalin Kulatilaka, *Real Options* (Oxford University Press, 1998).

3

The Molecules of Automation: Algorithms

It is a beautiful, hot summer day here in Houston, Texas, where I live. I find myself driving along the highway to get to work. My brain is ticking through the lengthy "to do" list in store for me when I arrive. I am completely oblivious to my driving—although I consider myself a careful, safe driver. What is going on?

A set of rules are running in my head that I barely notice. It includes rules such as my choice of lanes (I tend to like to drive the inside lane, but if that slows relative to the next lane over I might change). It dictates how close I get to the car in front of me. It guides me in my reaction to a stalled vehicle on the shoulder. And yes, it even has rules for how I react to the driver who rudely cuts me off!

Algorithms are goal-oriented, cascaded sets of rules that are influenced by external data, triggered by an event or constantly looping, never-ending. They can be incredibly simple or mind-bogglingly complex, but here is an important point: algorithms do *not* have to be complex in order to exhibit complex, sophisticated behavior. Stephen Wolfram in his book *A New Kind of Science* showed in his experiments on elementary cellular automata (very simple computational devices that simply flip binary bits of data between 0 and 1) that a simple rule applied over and over again can generate elaborate designs that appear to have come from a very sophisticated creator.

Dr. Wolfram gives us a great deal of hope here: we can design relatively simple algorithms that perform sophisticated tasks, meaning that the astonishing complexity of human decision-making does not necessarily have to be matched with equally astonishing and possibly intractable algorithms.

© The Author(s) 2019
G. E. Danner, *The Executive's How-To Guide to Automation*,
https://doi.org/10.1007/978-3-319-99789-6_3

If there is one principle that I would like you as a reader to understand, it is this one.

A fundamental feature of automation in this day and age is our ability to take algorithms as they exist naturally in brain cells and reproduce them, albeit imperfectly, in software. My driving algorithm is a classic example, and in fact lots of engineers are hard at work as we speak to capture this, *and in fact improve on it*—such automatic braking for objects in the road. It seems so odd that we can teach driving to a child over the course of a few weeks and months but it takes person-centuries to write code to do the same thing (this has to do with how the human brain learns which we will discuss in later chapters).

Here is a toy example of an algorithm to calculate the number of golf balls that fit into a school bus [1]:

Variables:

vGolfball	volume of a golf ball
rGolfball	radius of a golf ball
vSchoolbus	volume of a school bus
hSchoolbus	height of a school bus
lSchoolbus	length of a school bus
wSchoolbus	width of a school bus
p	packing factor of golf balls in space
nGolfballs	number of golf balls in the school bus

Logic:

```
        1.      Calculate the volume of a golf ball
rGolfball = 0.84
vGolfball = (4 x Pi x rGolfball)/3 = 2.5 cubic inches
        2.      Calculate the volume of a school bus
hSchoolbus = 78 inches
lSchoolbus = 30 feet
wSchoolbus = 96 inches
vSchoolbus = hSchoolbus x lSchoolbus x wSchoolbus = 1600 cubic feet
        3.      Determine the number of golf balls that can fit
                inside the school bus given the packing factor
p = 0.60
nGolfballs = p x (vSchoolbus/vGolfball) = about 650,000
```

Now you are well on your way to impress guests at dinner parties with your newfound knowledge of the intersection of golf balls and school buses.

Let's break down what just happened there. Making algorithms means starting from a blank sheet—the world is what you make of it and no presupposition is in place. Therefore, the first thing we do is to create for ourselves a list of variables—placeholders for data and information—for us to use in our algorithm. The variable list at the very beginning sets out the actors on the stage, and the scene upon which they will act.

In actual fact variables are the second element of the algorithm. We already mentioned one of the most important facets, the goal. *We want to find out how many golf balls fit into a school bus.* The goal is just as much a part of the algorithm as is the logic and should be treated as the first-class citizen that it is, by including it prominently in the listing.

In the previous chapter we put considerable emphasis on the value of the **hypothesis** for automation. The hypothesis should communicate the goal of the automation in some way. You can see here that I could have said, "I want to know the *exact* number of golf balls that fit into a school bus model X32E", or "I want to win the golf ball/school bus guessing contest at the county fair", each of which would change the algorithm in subtle but important ways. Resist the temptation to jump into automation immediately without first building a clear, concise hypothesis to govern the work.

In the Logic section, this particular algorithm has three steps, each with a title that describes what will happen. This is the 3-act play script by which the actors will behave. The stepwise equations are written in a fashion that anyone with even modest experience in writing software would understand immediately. The notion of blending both descriptive text and executable logic is critically important, because the goal of expressing algorithms in a format like this is communication to a wide variety of audience members— programmers, subject matter experts (SME), company leadership, business partners, analysts. Anything we can do to make the written expression communicate better to the consumers of the algorithm, the more effective our work becomes. In later chapters we will see how to bring the language of diagrams into our mix of expressive tools.

Notice that I incorporated an example (the data) right into the logic. Many times, the algorithm is presented in pure heuristic form—just the steps, ma'am—and the example, with the numbers, is computed later and separately. Here for brevity I simply rolled it all together.

This language is known as *pseudocode*. It isn't code that can simply be dropped into a computer to run immediately, but it is close enough to a code format for talented programmers to work with easily. At the same time there is enough pedestrian packaging with the comments and other explanatory text for normal humans to understand. If you are planning to become a participant in this new age of automation you do not have to be a programmer, however, you do need the skills to create algorithms in pseudocode such as the above. There is no standard for pseudocode, you can't take an online class in it. You simply have to think in systems and use that thinking to describe how systems work in these little granular bits we are calling algorithms (regardless of your talents it never hurts to take a brief course in any random programming language to get your mind ready to write pseudocode—many are available for free online).

Most automation consists of strands of algorithms chained together to perform a sophisticated task. The work of designing automation equates to the work of designing the algorithms that underlie automated systems. Implementing automation is usually very programming-intensive, but this comes only after someone has done the crucial step of building a thoughtful design. Pseudocode is the door that opens automation up to non-programmers, particularly to business people who simply want to create better companies.

One of the most famous yet ingeniously simple algorithms that gave rise to an equally famous company is Google's PageRank. Stanford University students Larry Page and Sergey Brin developed PageRank in 1996, looking for ways to create a new, more effective search engine for the World Wide Web. The original 1998 paper published by the founders describes PageRank in just a little over 200 written words [2].

In 1978 the Airline Deregulation Act was signed into law which changed the airline industry forever. Hundreds of new air carriers burst onto the scene offering low fares and new routes. It was around this time that a small group of scientists within American Airlines began to work on a set of pricing formulas that came to be known as yield management. Since then, yield management, often referred to as revenue management, has spread to hotels, car rental agencies, sporting events, and a host of other dynamic pricing applications.

Let's say you are a traveling salesperson. Your job is to visit 12 cities across the United States by car. The cities are spread fairly evenly between the East and West coasts with your starting point in New York City. Your employer has insisted that you travel in a way that minimizes your overall travel distance. What route—in order of the cities—do you take?

What I have just described is the famous Traveling Salesman Problem (TSP) which has been used as an archetypal challenge to explore many different kinds of fields, from computer science to supply chain optimization to drug development. TSP is a deceptively complicated problem. The number of solutions to TSP, including both the good solutions and the bad ones, equals the number of cities factorial, in our case 12! = about 479 million possibilities. It is clear that starting from New York on the East Coast, and then driving to Los Angeles on the West Coast, then driving back to Washington, DC on the East Coast and so on is not likely the best solution to the problem. Is there an algorithm that could help me here?

In classic optimization problems such as TSP there is one best answer (there actually could be multiple best answers in the case of a tie for first place). But let's say that we are willing to compromise on a complicated algorithm to get the one best answer in lieu of a "good" solution that is on average better than a random ordering of cities, but only if we can get to that good solution with a minimum of effort. Enter the *nearest neighbor*.

Nearest neighbor simply means that from my starting point I iteratively choose the nearest unvisited city until all the cities have been visited. Given the even distribution of the cities geographically, this is very likely to yield a better-than-random solution and in fact might be far better than an average solution. It certainly isn't perfect, but the economy of such a dead simple algorithm makes the choice of nearest neighbor feasible in many cases [3, 4].

Nearest neighbor is also referred to as a *heuristic*, a set of rules that short-cuts the brute force work of calculating every possible solution to a vast problem. Humans use simple heuristics frequently in their everyday lives, often without even realizing it.

Consider the lowly ant. Their brains are tiny. A single ant can only perform the simplest of instructions, but it can do so nearly 24/7 across its brief lifetime. Yet, colonies of ants (and many other insects) perform highly sophisticated feats of engineering and design in everything from nest construction to defending against predators. Algorithms make this possible.

When ants seek a new food source, they randomly scatter in many directions. Upon finding food an ant returns to the nest while excreting a chemical called pheromone, leaving a trail of pheromone between the food and the nest. However, pheromone evaporates over time, so the food source that is closest will have the strongest dose of pheromone. Other ants are "programmed" to find the strongest pheromone path and reinforce it with more pheromone. Over time the closest food source is located, and all ants gravitate toward it. And here's the simple beauty of the algorithm: because it runs continuously, it is *self-healing*. So, if you were to put an obstruction between the ants and a food source, move away, then return to it several minutes later you will find that the ants have ingeniously built a pathway around that obstruction to the source—or, to another food source that is closer than the newly obstructed path. This kind of collective intelligence is the clearest illustration of the power of very simple, elementary rules followed by a large number of agents that results in sophisticated overall behavior, sometimes referred to as emergent behavior.

There are countless examples of simple algorithms in nature. So much so that this has given rise to the vibrant field of *biomimicry*: the science of imitating nature in systems that we harness to do work. In fact, the ant behavior in the previous illustration was used for just that. Ant colony optimization (ACO) is an optimization technique that has been used to optimize transportation logistics in a wide variety of settings. In my own work I have applied ACO to the problem of marine logistics—moving goods between shorebases and oil and gas platforms in the Gulf of Mexico via support vessels. Why ACO? Because no other methodology we had explored up to then had this clever self-healing feature, which is ever so important in marine logistics where weather events constantly interrupt your carefully laid optimal routes in mid-flight. Thank you, Mr. Ant.

Over and over we see that simple solutions can be very powerful, counterintuitively. Too often we assume that a sophisticated output in the face of a complex problem is derived from an equally sophisticated algorithm underneath. In fact, many of the technological assistants that come into play in our everyday lives are remarkably simple at their core. This is a fundamental principle of this book and is the reason for our repeated emphasis here. The simplicity coupled with the ingenuity of great algorithms is the reason the craft of designing algorithms is easily within reach for folks who are not professional programmers. In later chapters we will show you how to do just that.

So many valuable companies these days began, or were reborn, via an algorithm. Google started with PageRank; the fractional jet company NetJets, was derived from a mathematical model; American Airlines survived and in fact thrived through tumultuous deregulation with yield management; Oracle Corporation, the software giant, began from a 1970 research paper on relational databases using Cartesian algebra.

Here's the message: don't start a company, create a unique, clever algorithm and then build a company around the algorithm. If your company is successful, you gain the advantage of having your core intellectual property be both patent-protected and portable to other companies and industry applications. In later chapters we will show you how to create a "sandbox" or factory for producing and testing algorithms.

Automating existing companies is a different process.

This often means finding the subtle, hidden algorithms that exist in human brains among SME in the organization. Say the engineer who cares for that top-producing chemical plant, or the finance director who is a master at tuning the capital structure of the firm, or even the warehouse manager who can find inventory where no one else can. All of these talented individuals doing their jobs every day is what makes high performing organizations work the way that they do.

Once these SMEs are "found", the algorithms they use in their brains are reproduced in a diagram, then on to pseudocode, then real software. The software is tested against the SME on similar problems and "bugs" are corrected until the software is a faithful (enough) replica of that SME's thinking. Voila!—we've now created something called a *Digital Twin* from a living, breathing human! In Chapter 4 we will describe the delicate art of dealing with SMEs to gain the best results so that our Digital Twins are as good or better than their organic counterparts.

The next decade will be a journey across scores of human SMEs in industry after industry discovering and reproducing the deep ceded rules by which humans perform work. The public will be continuously surprised by the kinds of automation that can be effectively created—in areas from the law to medicine to architecture to engineering and design. In 1950 the pioneering mathematician Alan Turing first proposed the *Turing Test*, in which a human subject stood behind a curtain and asked questions of a computer or a human on the other side [5]. If the questioner could not distinguish between a human and a computer providing the responses, then, he said, we have definitive proof that "machines can think". There is no question that most of the automation considered for practical application today would pass the Turing Test, even when the questioner on the other side of the curtain is a highly trained expert in a very specific field.

Once we have accomplished the making of Digital Twins from human SMEs, such work will be faster and more consistent than ever before, freeing up humans from mechanical tasks to pursue ever more creative and innovative thinking—something, it seems to me, that we never have time to do in today's world.

Summary

Algorithms are the magic dust of automation. They get their start in life as pseudocode, written by someone with a goal in mind. Algorithms show the "what" and the "how" of the system well before any technology is brought to bear on automation implementation. Simple algorithms can be very powerful especially when performed over and over again. Ordinary business folk are just as welcome (and able) to design automation as are technologists, because of this leverage.

Part I of this book introduced you to automation and described a future punctuated by automated systems that are nearly inconceivable for us today. We unfolded the underlying mechanics of automation through algorithms. Now it is time to get our hands dirty. Part II is about the Nuts and Bolts of automation, an exciting "how to" side of the concept that gets us working and building functioning systems.

Bibliography

1. http://www.wolframalpha.com/input/?i=how+many+golf+balls+fit+in+a+ school+bus.
2. S. Brin and L. Page, 1998, "The Anatomy of a Large-Scale Hypertextual Web Search Engine", *Computer Networks and ISDN Systems* 30: 107–117.
3. https://en.wikipedia.org/wiki/Nearest_neighbor_search.
4. "Universal Method to Sort Complex Information Found", *Quanta Magazine*, August 13, 2018.
5. https://en.wikipedia.org/wiki/Turing_test.

Part II

(Digital) Nuts and Bolts

As an author, I read lots of books myself. I've drawn inspiration from the high-minded concepts elegantly described by noted authors. Good writing about the business of business is everywhere. Stimulation that comes from new ideas is fun and rewarding.

Yet, many times I have been left with the empty feeling of not knowing what to do with my newfound knowledge on the Monday morning when I return to work. How is my work life different now? Sadly, most of the time the answer is … not much different.

Let's change that tune.

This part will take us into the hard-and-fast application of automation principles in a way that leaves you with specific actions to take to make your companies better, stronger, more resilient. In equal parts methodology and technology, we will show how companies can build automated systems from scratch, or to take existing manual systems and make them degrees more automated. Here is where the proverbial rubber meets the road.

Put on your safety glasses and lets get started.

4

What Is Automat-able and What Is Not?

If you have made it this far I have likely convinced you that automation is here to stay, and that you can now choose to play a pivotal role in the coming wave. You have a sense of what algorithms are, and how they fit into an automation scheme, either for a new, start-up company or for a legacy organization.

So…where do we start? We follow the path of least resistance—putting automation where it makes the most sense and gives us the highest value.

To begin, let us address head-on the question posed by the chapter title: What Is Automat-able and What Is Not? To do that we need to set the stage by going back to the origins of automation.

No one can say for sure when automation began, but certainly, the Industrial Revolution unarguably is an inflection point in the history of the clever use of machines to do meaningful work. The perfection of the steam engine and its commercial adoption in the late 1700s began a wave of mechanization of the basic production and transportation of nearly every physical good in the economy of the time. Electric power kept the wave going through the late 1800s, followed by assembly line production in the early 1900s. The postwar period from 1950 onwards brought logic computing to practical application, followed by the popularization of the personal computer in the late 1970s and early 1980s. Spreadsheet software followed, setting off an explosion in our ability to create and analyze data at the individual level—cheap, easy-to-use technology at the fingertips of any business professional.

© The Author(s) 2019
G. E. Danner, *The Executive's How-To Guide to Automation*,
https://doi.org/10.1007/978-3-319-99789-6_4

At every single step the process was the same: build machines to gradually take the place of manual human effort, or to leverage humans to accomplish more output per person than before. So what is different about today? Why is automation any more fundamental or important than it has always been through the last 250 years?

The difference is AI–Artificial Intelligence. When AI was first introduced in practical form it showed us that computers were not relegated to rote repeatable processes alone, like calculating a number. Rather, computers could be employed to engage in logic and reason in similar ways to human decision-making. In fact, the early work in AI was precisely that—build a machine to replicate the human brain. We called them *expert systems,* which were really just dressed-up inference networks. Fast forward to today and we have advances like *Affective Computing*, which seeks to impart computers with the ability to both express and understand human states, such as emotion and empathy.

Automation was defined by its common use—in the 1970s and 1980s that meant robots on an assembly line. Even today many business people have this same mental model of automation. But the modern definition of automation includes cognitive tasks like diagnosing a disease or deciding which stock to buy on the market. When you combine AI with Natural Language Processing (NLP) we now have smart systems that can easily interact with humans audibly or via text exchange. Given the state of the technology tapestry today, the answer to the core question that we asked in the chapter title is this: *every product of labor* is automatable. Some things are harder than others to be sure, but in principle, given infinite time and resources we could automate any cognitive task on the planet.

Automation is not a solution looking for a problem to solve. Rather it is a technique for rejuvenating unknowingly underperforming organizations in the absence of an automation component. It is surgery with a scalpel, not painting with a roller. Done right, in just the right doses, automation can have a powerful, transformative effect on the organization. Therefore our choices as to which things we automate first in a journey toward full and complete automation are critical to the success of the program. In this chapter, I will offer some guidance to help you make good initial choices.

Let me start by telling you a personal story of my own encounter with this decision. I was hired by a very large retailer to help improve the forecasting of sales for newly opened stores. It turns out that the ability of the company to forecast first-year sales was lackluster at best—teams got it right about as many times as it got it way wrong. Surely, they reasoned, their mathematical forecasting methodology was at fault for the inconsistencies.

This was not a trivial issue for the retailer. First-year sales set the tone for the store, and decisions such as staffing, store size, and the supply chain configuration flowed from the careful understanding of sales volume. Moreover, the ever-so-important task of providing Wall Street with accurate revenue guidance year-on-year, quarter upon quarter dictated that we get these numbers right, or severe penalties would swiftly follow.

Like a detective surveying a crime scene I conducted a thorough investigation of the models and data sources they were using. It turns out that there was nothing wrong with the model, but it became clear that the *location* of any brand-new store had an overwhelmingly strong effect on first-year sales. If the store was located in a "just right" trade area with little competition, attractive demographics, easy traffic access, and high visibility, and no cannibalization from nearby company stores, the results were strong. Choosing a location just one city block away from optimal could make a huge difference. Get *any* one of these drivers working against you, and the results were all over the map. The model's methodology wasn't the issue, something much more fundamental was going on here.

So the central question now moved to: how does the company decide where to locate new stores? It turns out that a group of 6 to 8 experts makes these decisions. I got to know them well over the course of my work with the company. These folks were walking encyclopedias of local knowledge about the 60 key metropolitan markets in which they operated. One person told me things about my own neighborhood ("there's a new school going up on Maple Avenue next year") that I didn't even know about.

The process for choosing a location was this: the team would gather in a conference room and review the 10–12 store location candidates for that month. Each specialist would present to the team his or her recommendation for the locations in their assigned region. If the logic made sense, the team would "approve" the location subject to CFO final approval. Verbal challenges were common but outright rejections of the recommendations were rare. Looking from the outside it was clear that groupthink and confirmation bias had crept its way into the store location process, preventing these great experts from critically stress-testing their decision-making. My recommendation, which was finally accepted after much debate and resistance, was to build a model showing the implications, as measured by a sales forecast, of a store location decision, numerically accounting for all of the key factors we had listed before.

In effect my job was to recreate the thought processes of a human expert in the form of a computer model. I started by asking them a myriad of questions about how they think about a store location and evolve a decision.

Oddly this was unnatural to them—experts are heralded as such for their intrinsic knowledge, *not talking about or writing down their knowledge*. Like professional athletes they don't think about what they do so well, they just do it!

Over time I was able to construct in a series of diagrams the algorithms so deeply buried in their brains as to how to compute a new store location recommendation. In fact, it was a handful of subtly different algorithms, as each expert had their own take on the problem (this turned out to be one of the most gratifying "aha" moments of the whole project).

When most lay people think of AI, they think of fancy and exotic software like Google Tensorflow or IBM's Watson, using an unseen cauldron of magic technology potions that ordinary business folk cannot begin to contemplate. Fair enough, but as we said in earlier chapters AI by definition is simply re-creating human logic and reasoning in a software form. The diagrams here, capturing in a concise form the experts thinking processes about store locations is just as legitimate a member of the AI class as are these sophisticated tools. In the end, the secret sauce for the retailer, the invaluable intellectual property of the firm, boiled down to several dozens of rules that could be expressed on one large sheet of paper. Billions of dollars in institutional value right there in black and white.

I built the store location model in software using the diagrams as my blueprint. Once complete, the model's user interface showed a geographic map of the proposed location. Click and drag the store five miles to the east in the computer—a new forecast was calculated. Click and delete (close) an adjacent store—a new forecast again. What I provided to them was an infinite canvas for trying and trying again different options for store location until the optimal location had revealed itself. The teams began to use the models right alongside their own human expertise, as if another virtual expert (a Digital Twin?) was present in the room, forwarding its own votes. The team didn't always go with the model's recommendations, but it did provide guardrails around certain decisions that protected them against those "way wrong" decisions that might go undetected otherwise. In this sense, we automated a portion of a critical strategic decision and reduced the number of poor location decisions by an order or magnitude.

But wait, is this store location model really automation? Humans are still making the final call on where to locate stores—you are just providing them analysis. What is automated here?

This example is a special kind of automation called Human-in-the-Loop, or HIL. It is very common in this day and age for a human to make the final call in lots of key corporate decisions where accountability for a performance outcome is still an important organizational feature. That means that humans do make the final decision with the aid of algorithms and stand by the results. It also implies that those same humans are motivated to tune and refine the models when they stray from optimal performance. And finally, there is leverage—the same group of humans should be able to do more with the same effort with a HIL model than under human power alone, and in fact, this was a surprising benefit to the retailer's team I mentioned before. Because the debate was shortened by having this omnipotent model weigh in, they were able to consider more candidate store openings per meeting than before—about double.

This example is a good one because it highlights the kinds of criteria we should be using generally to decide when and where we automate, and in what dosage levels:

1. **Critical decisions by an expert or experts**. An absolutely vital decision that is made by a group of experts, one that means substantial value (or loss) if done well (or badly)
2. **There exists an algorithm, but not formalized**. "Mary has always made this decision in our company. She's been doing it for 30 years"
3. **Lead time to decision**. It takes individuals or teams a relatively long time to decide, often because the process leading up to the decision (data gathering, consensus building) is so time-consuming.
4. **Decisions made again and again**. Decisions that are made, like pricing, at regular intervals are good candidates for automation.
5. **The actual decision is impossible for a human**. Do you know how many possible combinations for seating 13 dinner guests around a table? Answer: 6.2 Billion. So how can we expect a human to decide which of 13 products to launch in what sequence to achieve optimal value?
6. **A need to eliminate bias and other human failings**. Things like *optimism bias* cause humans to under-forecast the duration of projects consistently.
7. **The technology equivalent is obvious**. A very large, sophisticated company I know employs an individual **full time** to take the results from one daily report and key the data into a spreadsheet for another analyst to use. Let's call him Joe. I have a feeling that many, many companies have legions of Joes littered across their organizations.

Now as you can see a key ingredient that made the retail store location model work was the willingness of SMEs to share their thought processes on making location decisions. Automation is both art and science, and I must say that dealing with SMEs is probably one of the highest art forms in the mix. Humans are forgetful, selfish, overly optimistic, biased, and fearful. It's the stuff that makes us human and we wouldn't have most of the amazing things in our human experience (love, art, music, joy, faith) if we didn't have this litany of "shortcomings". However, this certainly doesn't make it any easier for someone trying to extract something as ethereal as "knowledge" from another human expert and to fit that knowledge into a machine that is as annoyingly literal as literal can be. I have made every mistake that can be made in dealing with SMEs, so I've learned a few rules of thumb about this. We will cover working with SMEs in Chapter 7.

An often-overlooked form of automation in business is *self-organization*, which really does not involve a computer. Self-organization occurs when a group of humans follow a set of simple rules (sometimes call agent rules or ant rules) in order to achieve a collective goal.

Let's say for example that you wanted to fill a football stadium with fans in the shortest amount of time. One solution is random—have everyone descend on the stadium with general admission tickets in hand all fighting for the best seats. That solution is not only suboptimal, but also rather dangerous in practice! Another solution is tightly controlled: fans can only be seated when escorted by an usher. Much more orderly, but assuming there are only a few ushers per group of 1000 fans, this solution would take a very long time to complete the stated goal of filling the stadium.

With self-organization, fans themselves are to follow certain rules, like "when you encounter a wall, always turn left" or, "enter the stadium at the point closest to your seat", or "go to your deck level before searching for your seat" and so on. Studies have shown that the right rules applied to self-organization lead to near optimal results when followed. Moreover, you only need a few followers to get the crowd turning in the right way, exploiting the herd effect.

Self-organization arises in the real world in many cases with companies that need to marshal humans collectively, without the overhead of direct control. Simple rules as in the stadium example above take the place of direct control (the ushers), and in fact are superior to direct control because managers cannot be expected to hover anywhere and everywhere that a decision is made. If the ant rules are well known and followed, systems can work remarkably well on their own. Vehicular traffic, for all of its failings, is an excellent example of this.

Simulation is the mechanism by which we design agent rule-based systems [1]. The simulation model is the stage upon which actors are assembled. Running the simulation is like firing the starting gun, as the actors begin to act in accordance with the rules by which they have been programmed. The collective result, evolving over time, is observed by the modeler as the simulation runs. The modeler has the option to go back to the start of the simulation, tweak the rules, and re-run the whole system with the same actors to explore the effect of the rule changes on the results. The rule-observe-change loop is repeated over and over again until the modeler both understands the system behavior well under a wide variety of conditions, but also now knows which rule configuration achieves the optimal result.

A fundamental point to all of this is that simple rules, when executed over and over, can lead to counterintuitively elaborate outcomes. Recall the ant algorithm that we mentioned in the previous chapter. Any one agent (ant) has a very simple task to do, perhaps 2–3 trivially simple rules. But when these ants are put together in a colony, very complex functions like building nests and foraging for food can be accomplished. The same can be achieved in our organizations if the rules are carefully designed and clearly communicated. One theory of organization management puts forward the idea that the CEO is less a command-and-control figure than that of a "designer" of rules that allow a large distributed enterprise to perform well consistently [2].

A second fundamental mechanism behind self-organizing systems is the feedback loop. Think about driving your car down a straight road. You, the driver, observe the position of the car in the lane. You constantly adjust the steering wheel slightly to keep the car in the center of the lane. The feedback of position coupled with the movement of the steering wheel forms a control "loop" that keeps the car in the generally desired location with ever-so-slight variations.

A modern company consists of countless numbers of individual feedback loops whether the employees recognize them or not. To make an organization more automated, we must expose and formalize these feedback loops through diagrams, then test them for efficacy, efficiency, and accuracy. Time delays between action (car moving out of the lane) and reaction (adjust the steering wheel) can result in harmful and chaotic organizations that constantly waffle back and forth between undesirable states. In my own career observing both great organizations and troubled ones, I often find that the source of poor or excellent performance lies in the quality and speed of the feedback loops that govern key decisions. In later chapters, we will learn about the science of feedback in the form of a well-evolved methodology aptly called Systems Thinking.

A question I get at my public speeches about automation is this: can automation itself be automated? Often what people are trying to express here is the idea that automation could more-or-less spontaneously arise from machine observation, as opposed to the hard, manual work of diagramming and coding to achieve the same result. Can machines simply start learning and then take over the task once they've learned?

The challenge here is that this is not just learning, in the way that machines are tasked specifically to learn about one thing—the correlation between the conditions of a sports game and a prediction as to whether the home team or the visiting team will win, for example. For automation to be automated you almost have to have a universal learning machine–a machine that learns how to learn, so to speak. A machine that "figures out" how its microworld works by keen observation and then moves to learn the concepts that are important without an predisposition as to what those things are.

What we have described here is something known as Artificial General Intelligence (AGI), which is a hot research topic these days. Practical examples are few and far between at the moment, but given all of the energy spent on the field, I expect this to bear fruit in a few years' time. When it does work, I can envision AGI machines set up to observe the work of companies and reverse engineering the algorithms that make those companies run, even as the company experts are blithely unaware of the precise nature of these algorithms. AGIs might suggest better ways to run medical clinics or hotels or factories by simple observation. The science of automating automation using AGI is just around the corner. As a practitioner you would be well served by staying on top of developments in AGI.

Summary

The Chapter began with a question: What Is Automat-able and What Is Not? We have arrived at the answer: everything is automatable, including highly cognitive processes involving judgment and reason. But some processes are more automatable than others and these are the better starting points. You are now equipped with enough criteria to sort through the candidates for automation and skillfully choose the best ones to begin with, thereby building momentum for the later, more difficult children.

Having elevated the candidates, now comes the real work: building diagrams. At first glance you might think that making diagrams is child's play—anyone can do that. But that would be ignoring the myriad of subtle rules that go into making a *great* diagram, an effective one. The next chapter will raise your game with respect to diagramming systems for automation.

Bibliography

1. Mitchel Resnick, *Turtles, Termites and Traffic Jams* (A Bradford Book, January 1997).
2. Donald Sull, Kathleen M. Eisenhardt, *Simple Rules* (Mariner Books, April 2016).

5

Diagrammatic Decomposition of Corporate Functions

Like visiting a new country, automation compels us to learn a new language in order to converse among the citizens of a highly automated world. That language is diagrams, the lingua franca of systems thinking. The sooner you learn how to be fluent, the sooner you will be building algorithms for automating systems with abundant economic value. And like languages, there is a proper grammar to use so that your points come across clearly without misunderstanding. This chapter will show you how to diagram in a manner that maximizes your communication quality and ensures that the systems that are built from the diagrams are an accurate replica of the conversations that initiated the automation goal.

The most common application of diagramming here is when we find ourselves confronted with an existing legacy system that is not automated but could be, perhaps identified using the criteria we discussed in the previous chapter. A system could be a corporate function, such as managing the daily decisions in a supply chain or pricing a vast array of products in line with market conditions. Legacy systems use armies of human experts to make and effect these decisions, moving at the maximum capacity of human effort. Our goal is to allow those same humans to make many more decisions, faster, consistently, all while relieving those same humans to do more useful creative work, like interpreting the meaning of market events or other disruptions.

The first and most important component of your automation story is the hypothesis, as we raised in Chapter 2. This should be the very first diagram you produce. A hypothesis is a clear restatement of the problem you are

© The Author(s) 2019
G. E. Danner, *The Executive's How-To Guide to Automation*,
https://doi.org/10.1007/978-3-319-99789-6_5

trying to solve with your automation layer. The word itself harkens back to The Scientific Method which has underpinned all great works of science for the last twenty-four centuries. There is a reason The is so durable—it works! By creating a hypothesis, you are putting a lasso around the problem in a way that guides the subsequent design to be true to its original and agreed intent, and at a granularity that is appropriate to the business problem that is solved. Far too many technology-related projects fail—not because of bad technology but rather because of open-ended excursions that have no clear end game nor an anchor to a real business problem. Resist the urge (and the external pressure) to simply "jump right in" to the project and take the time necessary to get a sound hypothesis that most everyone agrees with.

Often automation hypotheses contrast the before versus after as in: "here is the performance of our current system and here is the performance we expect from the automated system". A good hypothesis should frame the ensuing work in just this way—the automation that you do should now prove or disprove the expectation that has been set. The simpler the hypothesis the better. If you follow the design guidelines given in Chapter 2, you will increase your odds of getting the full benefits of a concise problem statement.

Let me give you an example, this time from the healthcare industry. One particular firm we knew was founded by a group of specialist doctors. These doctors felt that they could perform surgery far better and more efficiently if they controlled the surgical process and infrastructure rather than relying on a network of local hospitals. They pooled their personal investment capital to build a surgery center with a number of operating rooms. Like any startup, the patient load was small, but then grew as patients were attracted to the surgery center by a doctor they trusted.

Scheduling an operating room is a complex task. The room has to be capable for the procedure, and cleanup afterward could be simple or comprehensive based on the kind of procedure performed. Surgeries are not scheduled in a first-come, first-served fashion, but rather carefully sequenced to minimize the conversion effort from one to the next and to accommodate a certain duty load for the doctors.

The doctors group promoted a very capable nurse to the job of scheduling the operating rooms. We will call her Molly. Molly had in her head all of the rules for what procedures needed what equipment, what kind of sterilization protocol—the whole body of knowledge was in her head after nearly 30 years of procedures. She used a cloud-based scheduling system that presented to her the procedures to be done, by which doctor, and operating room had the right fit. The system maintained an active calendar for booked procedures. Molly was so good that the center became known for how efficient it was, which in turn attracted more referrals and patients. The center was growing substantially year-on-year.

Per operating room you had between 3 and 4 procedures to schedule in a given week, i.e. decide primarily which order to perform them in. Quick analysis of the patient loads showed that by the following year that would grow to between 6 and 7 procedures queued up per operating room. The scheduler had an intuitive sense that this was going to surpass her ability to schedule these procedures. She was right, the numbers 3, 4 and 6, 7 are significant in this case.

The number of combinations for arranging any sequence of individual items follows the formula $n!$ or n factorial: combinations = sum of $n * n-1 * n-2...$ and so on down to 1. For 4 items that's a harmless 24 possible combinations. It is likely that Molly's mental rules could rather easily cut out most of the 24 by apply simple constraints of "you never do this" and "you always do that". In doing so she reached the optimal point without even doing the math most of the time, which in turn led to the popularity (and profitability) of the surgery center. However, when you move to, say, 6 procedures the combinations balloon to 720! Molly's gut instincts were right, just two more procedures in the queue vastly outstrip her ability to apply her mental rules, no matter how good they were. Add another procedure to make it 7 and you get 5040 combinations. I could say here "you get the picture", but instead I'll show you in Fig. 5.1.

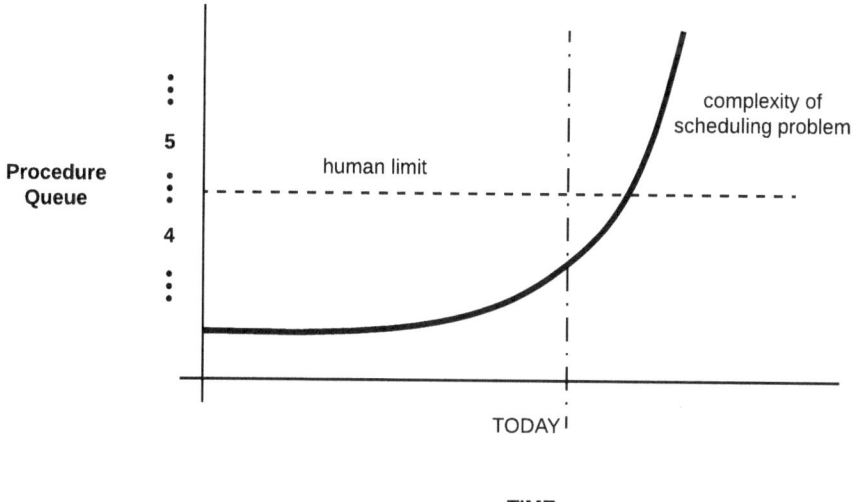

Fig. 5.1 Complexity of scheduling problem grows exponentially

This is the hypothesis for the problem statement I just mentioned. It is incredibly simple, deliberately so. But just that simple hypothesis encapsulates the reasoning, the motivation, and the intent of the automation: preserve the business capability that we are known and desired for, while handling the growth that we expect. It is precisely this kind of simple, crystal clear communication that we are striving for with our diagrams of the hypothesis.

With the hypothesis now in place, our goal is then to represent the way the *existing* system works in the form of a model—in this case a visual representation. Remember, the real system cannot be studied easily, so we are called upon to build an abstraction of the real system that is easier for a large group of human experts to digest, in the same way that a road map is far smaller and has much less detail than the actual land mass that you are traversing in your car.

Another subtle benefit of diagramming: language independence. While English is, in fact, the global standard for international business, not every audience or stakeholder will be equally adept at absorbing reams of English language text. In today's world where very few projects are exclusive to English speakers, the efficiency of communicating the intent of automation through diagrams is often an ingredient of success.

If you are the creative sort, by all means use that rare skill to build your diagram in a way that provides a clear and faithful representation of the system at hand that your audience (Subject Matter Experts and technologists alike) can understand.

On the other hand, if you have no idea how to start, let me suggest a format that has worked very well for us over the years of studying companies in all kinds of industries. It begins with the idea that all organizations are made up of functions, like "Compile Report" or "Build Product" and each function is comprised of smaller subfunctions and sub-subfunctions all the way down to their most fundamental unit of work. Each of these functions has inputs, and those inputs are converted to outputs. The input-to-output conversion is influenced by controls, much like how a recipe controls the baking of a cake. Finally, a function uses mechanisms as the infrastructure or tools to make the conversion. Hence, we have gone around the horn to create a guideline called the ICOM Definition: Inputs, Controls, Outputs, and Mechanisms. A function is represented by a rectangle, where each of these flows is positioned at key junction points: left, right, top, and bottom, respectively. Figure 5.2 shows how this works.

Diagrammatic decomposition then becomes the work of chaining together these functions in a way that shows how a company works in day-in-the-life format. Figure 5.3 shows a given page from a decomposition.

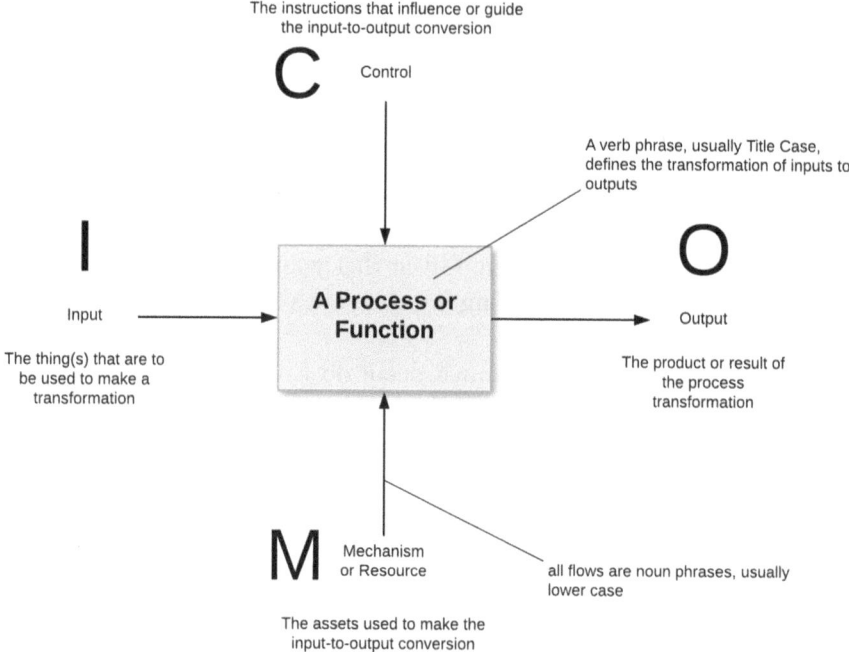

Fig. 5.2 ICOM function format

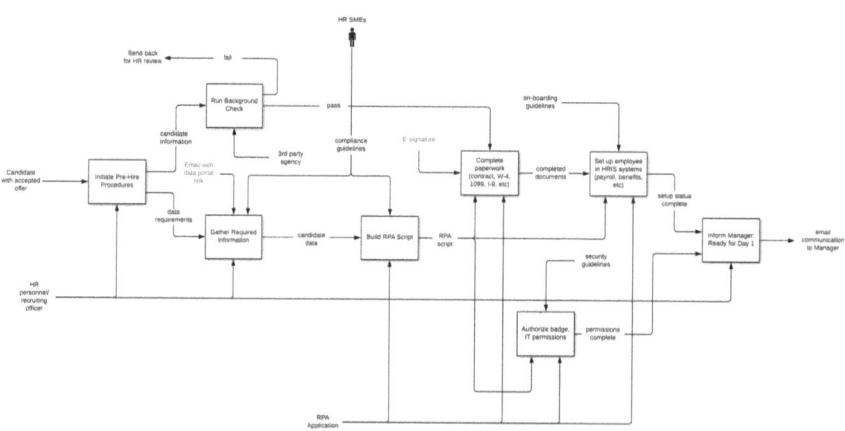

Fig. 5.3 Example diagram with functions chained together

A handy feature of this approach is the way that it supports hierarchy. A function expressed at a fairly high (parent) level, say "Sell Inventory", can be further decomposed into a number of child functions. In this way, a function can hold inside of it a whole new diagram that delves into the details behind the parent. Each of the children can be decomposed further in the same way, as can the children of children and so on as far down as you need to to make your faithful replica of the company. Hierarchy is helpful in two ways: first, it allows you the economy of expression, limiting any given page to a comfortably consumable amount of material, which is important to communication outside of your small team. Second, it allows the viewer to see the level of detail that they wish: from big picture to small detail and everywhere in between. Given that your automation project will have a broad audience, appealing to everyone's peculiar interest will be paramount.

I can hear you asking, "how much detail do I need then"? Here is where judgment will come as you go forth to automate all kinds of corporate functions. In the meantime, the hypothesis is your guide as to appropriate level of detail. Your goal is to represent the system just enough to answer the hypothesis question or assertion and go no further. You do not get extra points for making your diagrams more elaborate than they absolutely need to me.

Apart from hierarchy, layering is another crucial feature of this kind of diagramming. Layers imply that information exists on a diagram as several overlapping "panes of glass". A collection of objects may exist on a layer, but all of the layers are visible when looking at the diagram. Layers can be turned invisible, as if that particular pane of glass was removed from the deck.

Layering is very helpful in representing lots of information in one diagram. A common example is comments. Every diagram of an important system will carry with it notes that are associated with objects on the diagram. Perhaps these are comments from SMEs, questions to be answered, explanations, or notable features that need emphasis. While comments are important, they can also be distracting to the viewer who simply wishes to see the base diagram. If all of the comments are contained in a special "comments" layer, that layer can be turned on and off at the will of the viewer. Other drawing domains that are helpful when used with layering include exception conditions, future states, and phase or stage decompositions.

Both Hierarchy and Layering require drawing tools that support those features. In fact, the overarching message here is that one will need proper technical drawing applications to be effective and efficient with the creating of these critical drawings. Good tools are fit for purpose. Microsoft Powerpoint and even Microsoft Excel have been used to make drawings because they have rudimentary drawing tools. These are not recommended. Rather, if you are serious about making drawings with impact, drawings that communicate the essence of complex systems, you must acquire applications that are built specifically for this kind of work. Microsoft Visio is a good choice, as is Omnigraffle (Mac only). One tool that we particularly like is LucidChart, which is completely cloud based, and makes for easy collaboration with teams of far-flung stakeholders.

Often you will rely on Subject Matter Experts to aid in the content for the diagrams. At our company, we engage in a large number of what we call *whiteboard sessions* every year with our clients where we assemble the SMEs in a room for a guided conversation about a specific topic. Sometimes we draw out the diagram in real-time right before the eyes of the SMEs, sometimes we silently take notes and assemble the diagram later from these notes. Increasingly technology like Google's Jamboard and Microsoft Surface Hub make it practical to conduct these whiteboard sessions remotely with nearly the same efficiency as a face to face meeting.

Whiteboard sessions are where the social skills of the automation practitioner come into play—taking the ramblings of an expert and distilling it all down to the salient parts of the system that really matter to you (and are faithful to the hypothesis). There are no rules of thumb that I can offer here, as this is an acquired skill of artful listening. However, among my own staff we practice the skill by selecting random articles from newspapers and other business media. We then ask the student to build a drawing of the article. Who are the actors in the story? What are their relationships? Is there a sequence of events? Doing this time and again will build up your inner analyst and reinforce the practice of curating expert information.

We can take a tip from the world of software development here by adopting a few important principles from the Agile Methodology, a popular approach that developers and product managers use to efficiently create large software systems. One of the key tenets of Agile is the idea that we develop a little, then review, develop a little more, then review, repeated again and again in a circuitous path until the app is complete. You are inviting your reviewers (SMEs) to influence the design in increments as the system is built, as opposed to building a gigantic "spec" on paper then unleashing the developers to work from the spec in isolation for long periods of time. For our purposes as automation practitioners, this means that we build up our diagrammatic representations in small stages, sometimes in imperfect states as we attempt to get the characterization right. You must resist the temptation to spend an inordinate time with these diagrams to perfect them to the last detail before a review with an SME, anddispense with the notion that the SME finding a flaw in the diagram is a negative event. Rather, SMEs constantly correcting diagrams will draw them closer to the design, in a way that causes them to psychologically own the diagrams (this is an ideal outcome). In my own work, I have been known to deliberately introduce a logical error in my diagrams as a way to more actively engage an SME.

What makes a "good" diagram?

Like good art, experts know a good diagram when they see it, but describing good versus bad is challenging. Still, there are a number of guidelines to follow that will increase your odds of making a good diagram:

1. Use hierarchy, as explained above, to ensure that any one page is never so detailed that it becomes difficult for an audience to comprehend in one go.
2. Use landscape orientation except in rare circumstances where the information "wants" to be portrait.
3. Time and sequence ALWAYS go left to right. In other words, if function A generally executes before function B, then put A to the left of B on the page, never top to bottom except in the cases of representing a formal flow chart.
4. **Don't** use swim lanes for workflow diagrams.
5. Always **label functions and flows**; generally, functions will use verb phrases and flows will use noun phrases; verb phrases use Title Case while noun phrases use lower case.
6. Give each function block a 2 pica edge with a shadow.
7. Layer comments, exception conditions, and other elements that are distinct from the base diagrams.
8. Be very judicious on the use of color. 3 colors on the diagram palette should be the limit. The color red is special, and should only be used for very critical highlights; rather use a pastel-like color palette.

If you have followed the guidance above and have created a sensible, clear diagram of the existing system, then the real work begins. The diagram is the playing field for identifying automation opportunities. If the diagram is of good quality it will reveal clues as to where those opportunities lie, as shown in Fig. 5.4.

What about systems that don't yet exist? I'm building a startup company or a new division or a brand-new corporate function with no precedent. What then? The answer here is the same—you are still building a representation of the "thing" you are targeting, but without a current pattern to emulate. Rather, you build up your drawing from the intent of the business plan.

I know lots of examples of newly forming companies that wish to automate from their very inception. The business plan carefully spells out

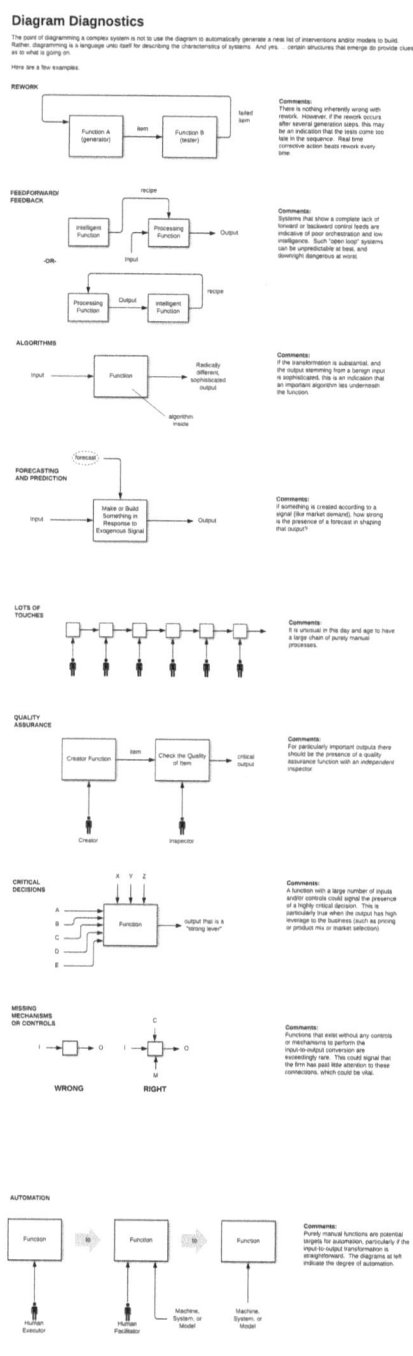

Fig. 5.4 Diagram diagnostics

the goal of the company. In effect, this is the Genesis of your hypothesis. Creating a diagram on how all of the company systems work stems from the ideas the founding team has on how the business runs. It is not as straightforward as automating an existing company, but the process is just as valid. You will tend to refine the diagram as the company evolves more than in the legacy case, and that is normal.

Diagramming in the manner that we have described here is THE most important skill for those charged with automating systems. That might seem odd to anyone first encountering automation, or even analytics. Most often when I speak to audiences about this subject, they commonly assume that programming and other technology mastery is the most important skillset. In actual fact it's the diagramming stage where most of the (permanent) mistakes are made, and conversely where the true value is created. The later technology build is generally a straightforward process that a good developer can execute with relative ease.

This serves to underscore what is one of the major tenets of this book, that automation is accessible to anyone with a keen understanding of how business works, not just the technological elite.

Summary

The very act of creating a diagram forces a process and a discipline on the investigation of a system. As you draw you think, as you think, you draw. Drawing anything from the memory of the original is contemplative work that slows down and focuses your thinking efficiently. No less than Leonardo Da Vinci showed us the power of observation backed by drawings with annotations of a breathtaking array of scientific and engineering disciplines from ornithology to hydraulics to optics. Automation to a degree is an art form because it can be implemented in an infinite variety of ways to accomplish the same goal. It is up to us the practitioners to become great artists, producing astonishing designs from a deep understanding of the world around us. Diagramming is the brushstroke of the modern era—learn it well and you can take your place among the leaders in the automation era to come.

A complete diagrammatic representation of a system becomes the blueprint for automation's design. The flesh on the bones we've just created is technology. It is time to think about the Art of the Possible in creating a new, automated version of the system we've just decomposed. The next chapter will provide a comprehensive scan of the technology landscape for automation.

6

The Technology Suite of Automation

Let us imagine that you have followed the practice outlined in the previous chapter and have created a whole series of objectively crafted diagrams that methodically tell the story of your system and the key points in that system that cry out for automation. The enabling mechanism for those automation opportunities is technology.

There is no one technology that produces automated systems, which is why you cannot, and never likely will, be able to "buy automation off the shelf". Rather the hard work of automation comes from assembling a suite of technologies in a carefully constructed architecture, with all of the underlying components well integrated. In this chapter we will explain a broad tapestry of technologies that are commonly employed for automation in a way that someone not formally trained in computing can easily grasp.

It is important for you as a practitioner of automation to understand these technologies at a "black box" level—what they do in principle, but short of the precise details of how they are programmed. In this manner, you will be able to assemble the constituent pieces of your automation solution into a cohesive structure, which we refer to here as an architecture. It would take several volumes of books to turn you into a practitioner of each of these technologies, so I encourage you to simply use this list as a jumping off point to conduct your own studies of learning the applications of the technologies as you see fit. In the meantime I will provide an adequate description of their role in the automation lexicon.

I will exercise considerable latitude in what I call "technology" in this chapter. Beyond computing technology there exist other *methodologies* that

© The Author(s) 2019
G. E. Danner, *The Executive's How-To Guide to Automation*,
https://doi.org/10.1007/978-3-319-99789-6_6

are exceedingly helpful in the process of putting automated systems together. In fact, they are so important that they carry equal weight to the computing side, and thereby belong in the same list. When I refer to the technology suite, I am including both computing and non-computing methods for understanding and communicating how complex systems work.

Technology is a vast landscape, constantly shifting. In essence, almost any technology can play a role in an automated system. I chose a list of technologies that are the primary building blocks of automation in the modern age, but I implore you to reevaluate technology frequently to understand some of the latest work.

Artificial Intelligence (AI)

AI brought us to this point. We would not likely be discussing the coming wave of automation without the advances in the technology of AI in the last five years. As the name implies, AI is a way to mimic the logic and reasoning of human brains in computing devices, especially when it comes to decision-making. In that sense AI could be something as simple as a flowchart with a handful of steps and branches, or a vast universe of knowledge, facts, rules, and inferences such as everything known about the treatment of a certain disease or the engineering of buildings. But no matter what the shape and scale of the domain you work with, the approach is the same: find out how humans think, document that knowledge in the clearest, most representative form, and recreate it in code. The first two will be covered in detail in the next chapter, while the last one—recreating knowledge in code—will be covered here.

In my view the tools for recreating human knowledge are the ones that allow you to move naturally from a simple concept like "revenues from product sales are equal to the number of units sold times the average sale price" to code in as few lines as possible and have those code lines be roughly recognizable even to the untrained eye. In a perfect world the business practitioner moves from the real world to pseudocode first, and the pseudocode rather easily maps to real code as created by an experienced software developer. There are two languages that are ideal for this purpose: Python and Wolfram Language. Python is an open source language, popular with the development community these days. It is elegant in its design and is supported by a very active community that is constantly creating free, open libraries that do everything from financial calculations to stylish visualizations. Wolfram Language was developed by Wolfram Research, the namesake of the brilliant scientist Stephen Wolfram. The flagship product Mathematica which uses the Wolfram Language as its programming paradigm is one of the most powerful and broad technical computing platforms available.

Natural Language Processing (NLP)

Automated systems interact with humans in real time. Really good systems allow humans to converse with systems—ask questions, execute commands—in ways that are "natural", i.e. the spoken or written word. When I ask the question, "What is the weather going to be like tomorrow?", I expect an answer that is sensible and allows me to prepare accordingly, even though you could state that same question in hundreds of variations.

NLP is more than simply consuming a written or typed phrase. Rather it resolves a phrase down into something called intent: "here is what the human is asking/talking about", no matter how the particular comment is phrased. That single, unequivocal intent can then be mapped in software to the particular body of knowledge that answers the question posed by the human. The process of creating a large body of knowledge whose underpinnings are vast number of intent-to-answer mappings is called *curation*. At present curation is mostly manual, using human SMEs and reference works, but much of the research in this area seeks to automate knowledge acquisition to the greatest extent possible. For an excellent free, publicly available example of NLP in action, try the knowledge engine on the web called Wolfram Alpha, available at www.wolframalpha.com.

Good NLP systems not only respond to human inquiries, but also exhibit two other important features. They are self-correcting, meaning that if a human receives a poor response from a given inquiry, a button is clicked by the user which in turn flags that question for later examination by human curators. These curators will then patch the holes in the system's knowledge for the next, similar inquiry by another user later on.

Second, good systems will *explain* their answers. How did the system formulate the answer step by step? What assumptions were used? Which experts put together this answer? Like the patient parent answering endless "but why?" questions from a curious toddler, NLP systems elaborate in great detail the path they took from the original question to the ultimate answer.

If you have studied how organizations work as long as I have, you develop a keen appreciation for the tremendous human effort that goes into collecting, sorting, collating, and presenting…information. Substantial increases in value are available to those organizations that automate this labor-intensive process. NLP features in that equation by serving up information on-demand to human questioners: an industrialized form of Apple's Siri, if you will. Wolfram Mathematica (from the same company that built Wolfram Alpha) and Stanford's CoreNLP library are two technology platforms available for implementing NLP solutions.

Systems Thinking

Systems Thinking is the science of systems, making a formal process of elaborating feedback loops that govern their behavior. It is said that structure drives behavior—Systems Thinking invites the automation practitioner to alter and refine the structure over and over to achieve meaningful, sustainable improvements in behavior or performance. If automation is an art, Systems Thinking provides the canvas, the brushes, and the palette of colored oils.

A system is simply a mechanism that performs work. At an atomic level, systems revolve around a series of feedback loops.

Let's return to my previous illustration of driving. The feedback loop at work here starts with a visual of the road ahead and the position of the car relative to the road. As the car drifts to the left, your brain registers the gap between the desired position on the road and the actual position: "I'm drifting 2 feet to the left of where I would like my car to be". That gap knowledge causes your brain to energize you hand on the steering wheel to turn it to the right. The car's steering linkages angle the tires in just right ways to move the car to the right. The eyes and the brain are constantly monitoring the new position until the car approaches the desired position, and the hand returns the wheel to the neutral position, perfectly straight ahead. We've just come full circle around the bend from view to gap to action and back again. This is a so-called balancing loop, where the system is designed to seek some kind of equilibrium balance. In the real world, sales, pricing, manufacturing, inventory management are all examples of balancing loops.

On the contrary, reinforcing loops characterize virtuous or vicious cycles that tend to move in one direction forever. A company loses market share to a competitor which in turn hurts revenues which then decreases their ability to invest in the product which causes market share to decline still further, and so on. We all know instinctively stories of reinforcing loops with restaurants that go out of business or neighborhoods that deteriorate, or on the flip side fashion trends that seem to spring up overnight.

Any given company is filled with a mixture of balancing and reinforcing loops. Systems Thinking is designed to help practitioners understand these loops as a means to understand the complicated nature of how firms behave. The fundamental unit of work in Systems Thinking is a diagram format called Causal Loop Diagrams. As the name implies, loops are literally drawn out on paper for everyone to see and understand.

You will often hear the term System Dynamics used alongside Systems Thinking. System Dynamics is actually subtly different from Systems Thinking. It is a simulation methodology whose purpose is to create a running computer model. System Dynamics brings loops to life by simulating their actions over time.

Systems Thinking is fundamental to all automation; its importance cannot be understated. Every organization—*every* one—at a control level is some set of both reinforcing and balancing loops working side by side. They are often invisible, hidden by the noise and chaos of the organization's daily activities. As an automation practitioner your job is to identify and elaborate the hidden systems in the form of a diagram, which could be a set of CLDs. Once done, System Dynamics can be used to test how the organization performs now, as well as how a revised organization might perform under a wide variety of conditions.

Machine Learning

In Chapter 3 we talked about generating algorithms from SMEs by carefully mimicking their moves step by step. There's another way to do this, not from the ground up, but by simply having a machine observe a system, human or otherwise, in action across many different variants of a task. This is called Machine Learning, and it is a fundamental technology behind Artificial Intelligence, and therefore is a pillar for automation as well.

Machine Learning is a special branch of AI but deserving of its own mention due to its importance in solving specific classes of problems. As the name implies, we use data to coax a special piece of software to learn the correlation between a set of inputs and an output. This data is aptly called a Training Set. Once trained, the Machine Learning model will then attempt to predict an unknown output given a set of inputs.

Here's how it works on a practical example, drawn from our actual experience in using Machine Learning. Let's say that you want to predict the outcome of an election by predicting which voters in a district are Conservative and which are Liberal. You are given 100 self-identified Conservative voters in that district and similarly you are also given 100 self-identified Liberals. Along with this for each voter you are given a handful of telltale attributes—things like income, level of education, number of children, residence post code, and so on. The 200 total "exemplars" form a training set in the sense that we are telling the machine: "here's a set of attributes, and this person is a Conservative; here's another set of attributes and this person is Liberal, and so on—200 times. Internally the machine learning software is forming a correlation, bit by bit, on how the attributes add up to the voter being characterized in one of two buckets. After training, we then grab a sample person whose attributes we know but whose political affiliation we do not, and "ask" the computer to make a prediction. Usually the output is in a format like this: "Judy is 80% likely to be Conservative and 20% likely to be Liberal".

There are many choices in Machine Learning software platforms, from high end commercial offerings like Tangent Works to open source libraries like Tensor Flow and SciKit Learn.

Optimization

Let's say you owned a furniture factory. This factory is capable of making three types of products: a chair, a table, and a desk. Each product consumes differing amounts of raw material, in this case wood. Each product also commands a different price and therefore a different profit margin per unit. Finally, the amount of human labor that goes into each product is varied as well.

On a given day I have so many workers, and so much wood. If I want to make the most profit on that day, what combination of tables, chairs, and desks should I make?

The answer is deceptively difficult given the vast number of possible sub-optimal solutions. With the same amount of wood and labor you could make, say, 200 chairs, 100 tables, and 50 desks, or, 150 chairs, 120 tables, and 75 desks, and so on.

Let me illustrate how vast the problem could be with a simple example. Say you had to arrange 13 different people around a table for a dinner party. You can arrange them in any order you like, but you want to make sure that Sally sits next to Ron and Marty does NOT sit next to Sue. How many possible combinations are there for seating 13 people around a table? The answer is an astounding 6.2 billion! Oh, and by the way, add one more guest for a total of 14 and the number becomes 87 billion. The solution space grows exponentially with each new incremental addition to the problem set.

Problems with extraordinarily large solution spaces are much more common than people think, largely due to the fact that we circumvent the problem by applying overly simplistic rules and satisfy ourselves with sub-optimal solutions. But in business, sub-optimal solutions leave money on the table. There has to be a better way.

Optimization models are computer algorithms designed to ply these vast solution spaces to find the optimal solution. They don't actually calculate every possible combination but rather take shortcuts using some clever techniques that have been improving since they were first introduced almost 70 years ago.

Every optimization model works from the same frame, which consists of 3 key components: an objective function which calculates the thing we are trying to maximize or minimize, a set of constraints that define what is a "legal" solution, and finally a set of variables that the model changes again and again to calculate a wide range of solutions.

Let's see how this works by returning to our factory example.

Our objective function is the goal, which we said at the outset is to maximize the profits of the day's output. We can calculate this by creating a formula that considers the revenue of any given product less its direct cost of labor and materials. This is in turn multiplied by the number of product units made on that day.

The constraints are represented here by labor and materials. I need so many worker hours for each product and so many units of wood for each product, and I cannot exceed the total capacity limitations. As a user I have to describe the constraints to the optimization model so that it knows the difference between a valid solution and an invalid one.

Finally, the variables are the levers that the model pulls and pushes to get a different value of the objective function. In this case the variables are the number of tables, chairs, and desks that can be made. You can think of the variables as similar to the dinner party example I presented above. The model will try an initial arrangement of the guests—no that's not good enough, shuffle, yes, that's a little better, shuffle again, ok that's a little better, shuffle again, no that's worse now...the model iterates across the solution space looking for the magic combination of tables, chairs, and desks that maximizes the profits on that day for the furniture factory owner.

In my experience working with an extensive collection of organizations over the years I see these optimization problems cropping up in every corner. In far too many cases we rely on humans to ply the solution space using simple rules that rarely change over time (even as new humans take over the role). Gradually the problem gets exponentially larger from the original but humans still make the call. More money left on the table.

Optimization models should take the place of humans for critical decisions like our stylized example with the furniture factory. Humans, in turn, should be tasked with the strategic questions like, "hey I notice when I add just 2 more workers my output goes up 20%" or "I can pretty much make the same products but with 10 less units of wood that ends up getting scrapped". Exploring the non-linear effects of constraints reveals inflection points in the solution that in turn generates value on top of the decision automation enabled by the optimization model. Once again, automation is simultaneously adding value to the organization while enriching the value of the human worker.

The technology behind optimization ranges from the crude engine inside of Microsoft Excel called Solver to a powerful evolutionary algorithm contained in Charlotte Software Systems' range of products.

Factory Physics

Every business at some level is a factory. *Every* business.

You may not think of a bank or insurance company or real estate firm as a factory, but if you think about the interior of these businesses, they have many of the same characteristics. They have raw materials (often information or fixed assets), queues, cycle times, work flows, inventories. They aren't making widgets but if you bend the physical production metaphor just a bit it is easy to see the underlying factory at work, even as the work products are invisible. By thinking of any business in this way, we can apply factory-like thinking to a broad range of businesses, which in turn makes them more amenable to automation.

In the past we had an opportunity to work with a very large European bank on a set of risk calculations. The bank needed help in systematizing a largely manual process that was fraught with error and inconsistency. Calculation of a risk metric for the portfolio of loans is one of the most crucial steps in a delicate system of balancing the bank's performance against the appetite for risk that is appropriate for that class of loan positions. As we broke the system apart we discovered a labyrinth of workstreams and sub-workstreams—the diagram completely filled two rather large whiteboards. Step 39 is completely manual, then the work product is handed over to System X to accomplish Step 40, and so on. You get the picture.

The bank had a factory here that produced an intermediate product called a risk metric. Actually there were many different colors and shapes of risk metrics, and these widgets further plugged into the bank's larger production complex in countless ways—generating new loan products, valuing the inventory of positions, and injecting overlapping capital into other business units. Our job was to make the risk factory more efficient.

Factory thinking seeks to weed out non value-added time, time consumed in process where the widget is not being processed. It also seeks to reduce the labor content of each step, cut down on parts that fail and have to be reworked, and shave work-in-process to an absolute minimum. All of these were foreign concepts to bank executives, but once exposed through the diagrams, factory thinking took over and unleashed a torrent of sensible improvements where none had existed before. Version 2.0 of the risk calculation factory for this bank resulted in fewer steps, fewer humans enabling those steps, greater accuracy, and smoother integration with downstream financial products.

For a deeper reference on the algorithms that govern factories, see the book *Factory Physics* by authors Wallace Hopp and Mark Spearman [1]. Many of the algorithms for factory-like concepts such as "inventory" can be reinterpreted for firms that do not consider themselves to be factories. By employing Factory Thinking, it is as if you have put on x-ray glasses that enable you to see company functions in a completely different way, which in turn inspires solutions that are not obvious to most insiders.

Visualization

Applied automation should not simply automate a function, *but it should show you how it does so*. It can if it incorporates visualization.

Like a TV chef who shows you each step in a recipe, automated systems can express themselves visually as the algorithm works through its logic. It is "explaining" how it does its work in real time, including what sensor readings are most pressing, what intermediate calculations are produced, and what trends it is tracking. The most efficient form of visualization is through graphics animation, artfully using color, shape, and scale to focus your eyes on the most important outcomes as time unfolds.

Dr. Edward Tufte is the most well-known and extensively cited expert in the field of information visualization, having written four landmark books on the subject over the last two decades [2]. His work will show you how (and how not to) take vast sets of data and make them meaningful through proper visualization techniques, born of the science of information design.

A rather extensive open source Javascript library called Data Driven Documents (D3) is the foundation for many browser-bound visualizations. The library contains an enormous number of widgets for expressing every kind of data form one could imagine.

Digital Twin

A fundamental pillar of automation is the Digital Twin, a computer-bound replica of a real asset, or perhaps a person. The Digital Twin emulates its real-world counterpart in software. A Digital Twin of, say, a machine used in a production process carries with it all the virtual behavior, the performance data, and the attributes associated with the asset. The concept extends to all kinds of assets from "hard" ones like products to "soft" ones like contracts. Pseudocode is often used as a stepping stone between the asset itself and a proper Digital Twin.

This is important in automation because we often automate the Digital Twin as a test before we begin the much more expensive and risky process of automating the real-world asset.

Graph Databases

By now as a professional you have encountered databases in your daily work. Even a spreadsheet is a mild form of a database, holding information in a structured way for easy retrieval. Perhaps your company has SAP or Oracle or some other enterprise business system for carrying out the functions of the back office, such as General Ledger and Procurement.

At a fundamental level the databases you have encountered thus far are based on a Cartesian relational model, a fancy term for describing data stored in tables, which are laid out in rows and columns. The rows represent many similar data elements, while the columns define the attributes of each data row. A table of customers would contain IDs, names, billing address, Tax ID, and so on. Relational models have been around for over 50 years, ever since mainframe computing came into being.

Let's say you have two tables related to each other through a key. One table lists all customers, while another table lists all orders, from every customer. As I grab hold of one particular customer in the Customer table, I use that firm's unique ID to move over to the Order table to find all of the orders for that customer only. As a user I can accomplish this by running a special query statement in Structured Query Language, or SQL for short. When an SQL query is executed to find the orders for a given customer, the database system uses a technique called a join to link the two tables in memory, then perform the extraction that the user wants. A join is one of many Cartesian logic operations that are performed in the background by the database system.

In the last 5 years we have seen the rise of a different paradigm for storing and retrieving data: the graph database [3]. As the name suggests, data is fundamentally represented as nodes and edges. George is a node and his barber Ned is a node also. George and Ned are connected to each other through an edge, and that edge from Ned to George is termed "is_barber_of". If I repeat all of the important relationships to George, I would have a dense network of nodes and edges, and it would in turn allow me to run queries against the data like, "how many 'friends_of' George also get their hair cut by Ned?

A node can be any object—a person, a contract, a production plant, or a product. Edges themselves can contain data. Returning to the George–Ned connection, that edge could hold information about the is_barber_of relationship between the two, such as the date that Ned first became the barber to George, how much he charges him, and the average frequency that George visits Ned.

Graph databases became popular just as we were changing the ways that we wanted to store information. The goal is to collect and store whole bodies of information, such as all known facts about Leonardo Da Vinci, or the elements of every constitution for all countries in the world. A universe of information like this is called a corpus, and the diverse structure that is required to store all known facts about something clearly transcends the rigid row/column format offered by relational databases. Querying graph databases involves a "traversal" of the relationships from one node to another through edge connections as opposed to the join mechanism in relational databases. It turns out that traversals are computationally faster and more efficient as well.

Automation systems must draw inferences from data sources in the relevant world around them. A system that is deciding about whether to offer a loan to an applicant or another that decides how to configure a food processing plant to make the most profitable products or a system that is running a supply chain in the midst of a natural disaster must retrieve all sorts of information to effect good decision-making. That information is not always gathered in orderly rows and columns. Graph databases make it possible to piece together a corpus that can yield a much richer set of facts and relationships that are required by more sophisticated cognitive applications.

Blockchain

Universal. Distributed. Ledger.

Put all transactions for every firm on one secure ledger system. This is the promise of Blockchain technology. As of this writing Blockchain is coming into its own as a rapidly maturing technology that is finding real applications across industries. It also happens to be one of the most over-hyped technologies out there, with an odd emphasis on cryptocurrencies which are simply one manifestation of "apps" on top of the blockchain infrastructure. Don't let this sideshow distract you. Blockchain has some substantial benefits and is useful for a certain class of automation.

A ledger is an organized, human readable record of something, much like my bank statement. Blockchain allows almost any electronic object to occupy space on the ledger, beyond words and numbers to include things like songs, engineering drawings, software, and contracts. The universal nature means it's the same shared utility everywhere. And the distributed part implies that many, many copies of the identical true ledger exist around the internet. Blockchain uses an ingenious, powerful scheme for maintaining the integrity of the contents of the ledger and preventing tampering.

Part of the magic of Blockchain is in the way information is securely exchanged ledger to ledger with a flexible permissioning system. If I want my doctor to see my vaccination records, but not my mortgage balance, I simply generate a special electronic key (a long sequence of numbers and letters) specific to my doctor *and* to the vaccinations. My doctor uses that key to unlock that very special portion of the blockchain to retrieve only the information for which I have granted my permission.

Naturally one of the first applications of Blockchain has come in supply chains. By uniting all of the actors in a given supply chain onto one ledger system you can do things that you could not do before, like analyze the horizontal efficiency of goods flow, or trace a defect back to its source. But even more promising is a feature of Blockchain technology known as the *smart contract*. A smart contract is actually a piece of code that maintains a position on the ledger right alongside the static transaction records. As that smart contract is referenced by a user, the terms of that contract are automatically executed by the Blockchain without any central authority needed in the loop.

Say you want to rent an apartment for 3 months in a given city. You and the landlord of the building jointly select a tailor-made app that forms the smart contract for renting the apartment. The app takes your application, checks your credit, collects your deposit, issues you an electronic key to your smartphone, and you move in. Every month the app sends you a rent request. At the end of the 3 months your key is disabled. All of these clauses in the contract can be executed automatically via a correctly written smart contract.

Now let's move the use case to a more industrial setting. Say you are a shipper of bulk materials like sand and gravel from one marine port to another. A load of sand in Wisconsin has to move down the river system to the port of New Orleans. You and the vessel owner have a smart contract that books and schedules the load on the appropriate ship. The vessel develops an engine failure? No problem, a smart contract can automatically transfer the load to the next available ship.

Blockchain will eventually become a public utility for business transactions large and small. As such it is an infrastructure perfectly suited for automation. The automation will come in two ways: smart contracts executing functions that used to require a human-powered central authority, and horizontal analysis of value chains on behalf of the merchants at the end of the chain.

IoT

The Internet of Things (IoT) is a widely documented technology layer that allows real-world sensors to inform an intelligent system. Certain features of the world—the temperature in a greenhouse, the presence of a human, or the speed of a vehicle are important to the decision-making processes that run in software and IoT forms the channel between the two. In most cases, sensors values are wired into a cloud-based system for secure, efficient retrieval.

One of the most active areas of development in IoT is the concept of sensor fusion that we mentioned in the very first chapter. Sensor fusion speaks to the idea of enlisting multiple sensor values as a means to inexpensively draw a particular conclusion about the real world. In the remarkable case of Amazon Go, sensor fusion allows its systems to identify a product lifted off the supermarket shelf using cameras and weight sensors, rather than an expensive RFID tag affixed to a $1.69 bottle of ketchup. More on this in later chapters.

3D Printing

Also known as additive manufacturing, 3D printing allows a machine to generate a physical object by directing the buildup of material in a manner similar to how a page printer uses a printhead to draw characters and images on paper in 2D. Because the process is controlled by a computer, almost any object, no matter how elaborate, that can be described to a computer can be 3D printed. Put several parts together and you can 3D print an entire machine. An amateur hobbyist friend of mine is using his 3D printer to print … a 3D printer.

There are some extraordinary implications of 3D printing, perhaps too numerous to list all of them here. But in terms of automation we could easily think of machines repairing themselves by printing a replacement to a broken part, or fashion designers in Paris sending a jewelry design to 3D printers in China at the click of a mouse.

Simulation

Automation without a deep understanding of the system targeted for automation is dangerous business. Simulation allows us to generate replicas of all kinds of systems in the form of a computer model, then try various changes to the system to observe the implications. A common use case for simulation in automation involves the modeling of the unautomated function, then testing various configurations or degrees of automation to understand how it will work. This is of course directly consistent with our introduction of the Digital Twin earlier in this chapter.

As exciting as the power of each technology is, things get even more interesting when any two of these technologies combine to build another foundation that is greater than the sum of its parts. As this book is being written, the blockchain community has reached out to those in the graph database realm and is hard at work looking into opportunities to combine the two in creative ways. A big part of your job as a practitioner is to monitor these developments often.

Summary

Technology is the enabler of automation, which comes about by selecting from among a large collection of individual technologies into a meaningful suite, with each performing a role in accordance with its functional strengths. Automation practitioners develop architectures that assemble the technologies together like the parts of a useful machine. It is important for practitioners to know how these technologies work and what they contribute, but it is not necessary for them to understand the minute details of how they perform "under the hood". This chapter provided a thorough illustration of some of the most significant technologies available for us today to use in automating systems, although the landscape is constantly changing.

Let us now illustrate the marshalling of a handful of these technologies into a realistic architecture for performing an automation function: building up an automaton from a diagrammatic blueprint.

Bibliography

1. Wallace J. Hopp and Mark L. Spearman, *Factory Physics* (Waveland Press, Inc; 3rd ed., August 2011).
2. Edward R. Tufte, *The Visual Display of Quantitative Information* (Graphics Press, May 2001).
3. https://neo4j.com/blog/7-ways-data-is-graph/.

7

Building an Automaton
from a Diagram Blueprint

We have now woven together the dual themes of diagrammatic decomposition with a sizable technology tapestry to bring us to the point where we are now ready to create a unit of automation. For our purposes we will refer to this as an automaton, a divisible subsystem that enables the automation of a corporate function. It is the work of practitioners to constantly wander the business landscape building automatons that plug into the enterprise in specific areas that yield a high-value result for the company, either through labor savings, increased speed, higher quality, enhanced agility, or some combination of these intrinsic benefits.

The automation process I recommend is shown in Fig. 7.1. The first step is the creation of the Hypothesis, which you learned back in Chapter 5. Next comes the creation of the Digital Twin from the physical system, spinning out function maps—natural algorithms—along the way. SMEs work closely with the practitioner to build the blueprint diagrams for the automation. Automation is then carefully simulated in stepwise fashion until the behavior is well understood, thus beginning a cycle of design-simulate-build actions to automate the target system by degree. At a level of automation that largely removes the humans, you will have options to completely redesign the system under its new, automated frame. In fact, many automation experts have suggested [1] that the real economic payback from automation only comes when that automation is combined with a complete rethinking of the old process, now newly automated. I agree with this thinking, but I'm more lenient in endorsing initial projects that automate without complete process redesign.

© The Author(s) 2019
G. E. Danner, *The Executive's How-To Guide to Automation*,
https://doi.org/10.1007/978-3-319-99789-6_7

The Automation Process for Large Scale Systems

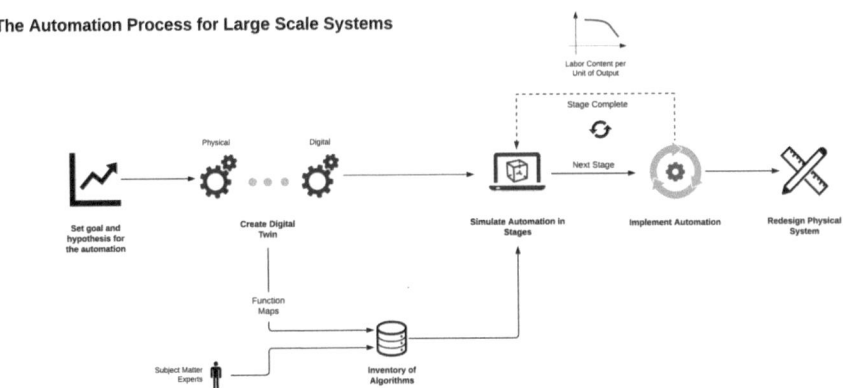

Fig. 7.1 The automation process

The best way to describe any sort of building process is through an example. The example I will describe to you is an abstract but realistic composite of several different actual projects we have undertaken as a company.

ACME is a chemical manufacturer. They produce a slate of base chemicals that go into a variety of plastic materials from raw materials like natural gas. Large-scale plants effect a chemical reaction with heat and energy to produce the chemicals in various grades at the highest physical yield possible. In fact, ACME's plants are highly automated and efficient, running elaborate process control computing systems that orchestrate pumps, valves, boilers, and other equipment in just the right sequence. ACME is proud of its sophistication and considers itself one of the leaders in automation, in an industry where automation is a common fixture.

But wait, there's another story here. These highly automated plants produce a finished product, and then that product is distributed to markets around the world. Upon further examination, the "distribution function" is highly manual, with a very high labor content per unit of finished goods relative to the manufacturing side. You ask to see the Distribution Department. There they are—an entire floor of people huddled in cubicles on the phone or tapping numbers into spreadsheets. The buzz of activity is palpable—everyone is extremely busy! Our first clue that something is amiss is the stark contrast of the plant, humming away with a small task force of humans tending to the machinery versus the Distribution Department which is an army of humans working desperately hard every day, every hour to find homes for the products. It would seem that the level of automation in one side of the firm does not appropriately match the level of automation in the other. If automation was good for manufacturing, and indeed expected in the industry, why couldn't the same automation apply to Distribution? In fact, could it also be that the lack of automation in Distribution hinders the benefits of the great work of automation one step upstream of it?

We sit down with someone in Distribution, in a role called Outbound Supervisor.

"What is your job?", we ask.

"My job is to take the finished goods inventory in the tank farm at the plant gate and set up shipments to customers. I work to get rid of the inventory so that we can make more product."

"OK, that makes sense", we say. "How does that work?"

"About half of our inventory volume is on long term contract, what we refer to as offtake agreements. We have to satisfy the contract which says that so much product every month has to ship to a certain location. The volume rarely changes month to month. The other half is ad hoc demand where we get to choose where to put the product from among a large number of demand points. In a given week we match 12 products in the slate to about 60 possible locations to come up with about 200 shipments by the combination of rail, inland barge, and pipeline. I build all of this in an Excel spreadsheet, and then hand it over to the Transportation Department to execute."

"How do you decide what demand point to go to with any given parcel, and which mode to use?"

"We try to figure that out given other shipments that might be heading that direction".

"Is there a big difference in cost of delivery and the price you receive at one location versus another?"

"Oh yes, especially during disruptions like when a competitor's plant is shut-down or natural events like floods".

"Do you ever trade off the net margin from one location to another to find the optimal overall best set of shipments to make the company most money?"

The gentleman looks at me warily, "Son, I've been doing this for twenty years. We try our best but we simply don't have the time to do all of that analysis. The best supervisors know how to stretch our dollars and they do a great job however, we can barely keep up with the demand as it is. At the end of the day we're happy just to get the product out of our tanks so that we can make room for the next round of production from the plant."

If we return to the criteria in Chapter 4 we see that in this short dialog with the Planner that many of those are satisfied here:

1. A human is making a critical decision: where to send the product to market.
2. "The best supervisors know…" implying that there is an algorithm in place used by the most experienced, knowledgeable personnel but there is no evidence that these rules of thumb are written down anywhere (otherwise all of the supervisors would be using them).
3. The problem is far too vast for a human. Twelve sources against 60 uses with 200 mappings between them is an astonishingly large solution space. Feasible solutions will number in the hundreds of thousands.
4. The same decision is made every week, week-in and week-out.
5. It takes too long to make a decision. In this case, there isn't enough time to make the optimal decision, so the company is satisfied with a less-than-optimal decision that is dumbed down to whatever a (stressed) human can accomplish in a given day.
6. The technology is rather obvious. At its core this is a classic optimization problem as we discussed in the previous chapter, oddly very similar to how the plant works today, just infused with an intelligent model.

If we were to automate the Distribution function, more products could be distributed with the same number of people. Moreover, the humans could be repurposed for more creative, thoughtful work. Remember what the Planner said about disruptions causing disturbances to the market? Why not get humans thinking about clever ways to exploit those disruptions, then springing into action when they occur? The name of the Distribution Department changes to *Market Optimization*, fitting with their new, more exciting role. Perhaps disruption actions in turn become an algorithm itself later on, triggered by the system "sensing" or predicting a disruption automatically? Humans then move on to the next most creative goal, and the cycle repeats.

What we have described here is a far more valuable (to the company) and rewarding (to the person) role for humans than simply repeating the same function over and over again. Automation in this case seeks to make life better, more enriching for humans while delivering top-notch value for the company at the same time. The whole zero-sum game (automation wins, humans lose) being presently portrayed in popular media is an unfortunate myth.

In Distribution, as with any other process, there are inputs and outputs. The output of one sub-function is the input to another and so on proceeding in sequence from left to right until the final output is realized. Getting started is simply following the trail of initial "ingredients" (typically source data and a goal) for the function through to the intermediate calculations then on to some reasonable definition of completion. Along the way you will define which elements control the input-to-output conversion as well as the mechanisms or resources necessary to enable the conversion. It is not rocket science, and if you are like me, you will try a drawing, throw it away, and try again countless times until you feel comfortable enough to review your result with a SME. The key is to avoid representing too much detail in any one page by cleverly using hierarchy as illustrated in a previous chapter.

Orient your thinking in small, divisible steps. What does the supervisor first assemble to get ready to make a weekly distribution decision? What is the first step? Where does the output of that first step go? How is that processed? It usually only takes a few of these mental questions to get into the right frame of mind to start diagramming. Function blocks are named using verb phrases like "Generate Forecast" or "Ship Product". Flows between the function blocks are labeled with noun phrases like "Orders" or "Summary Report".[1] Structure the progression of steps from left to right and in doing so, tell the story in a picture as you had practiced with news articles as was suggested earlier in this book.

When the draft is complete the next step is to review the draft with a human SME. There is an art to working with SMEs, coaxing them to give you just the right information you need, and no more. Keep this in mind: SMEs spent their careers doing what you've just put into a single page (or so they perceive). That won't sit well with most people, so they will often "invent" ways to make it more complex than you've shown. That comes via the introduction of myriad exception conditions that they've experienced over the years, and they are rarely shy about telling you lengthy tales about some part of their past. That is interesting data, but as practitioners we must ruthlessly apply the 80/20 rule to the information we receive, which requires us to tactfully steer the SME to answer the pertinent questions about the mainline operation, the exceptions to be addressed later.

Terminology matters. The words that SMEs use form the special language of the company, unique to the industry and the organization itself. As we are striving for quality of communication in our diagrams, the more we can use the "local language" of the business the better.

What I often find with new practitioners in automation is great detail in inputs and outputs, but much less, and even complete omissions of the

---

[1]If you find that the verb phrasing for functions and noun phrasing for flow lines is unnatural, this could be a clue that your diagram is not structured correctly.

controls and mechanisms. I believe that is because we've been engrained with process flow and workflow diagramming from the lean/six sigma world, which underemphasizes anything that isn't an input or an output. Automation is a different game from lean thinking, and therefore has a different focus. Controls and mechanisms are a vital part of the expression of cognitive corporate functions. For controls, think of it this way: what instructions, criteria, or considerations do I have at my disposal in making the input-to-output transformation? For mechanisms: what infrastructure, people and machines, systems, are necessary to enable the input-to-output conversion. It is very rare to have any corporate function that is free of all controls and mechanisms.

The mechanisms, collectively, are an especially powerful way to express the underlying machine architecture that is in place in the current system. Since automation involves the replacement of human action with machine action this is a vital playground for experimenting with a variety of automation options.

A completed diagram of the way the current system works is like having the patient lying on the operating table, chest open to expose all of the vital organs. A trained physician can simply look over the body and make skilled assertions about the ailment, aided by the data from other testing that had been performed pre-surgery. Your work is similar: a faithful replica of the Distribution function in our case here provides a wealth of clues as to opportunities for automation that would value for the ACME.

Your eye is drawn to a particular function called Create Parcel (see Fig. 7.2). Returning to the Outbound Supervisor, you ask some questions:

"Tell me about this Create Parcel function."
"That is where we unitize the inventory for shipment. We create parcels which are units to be shipped."

"How does that work?"
"There are certain parcel sizes by mode that are economic. If you are shipping by train, traincar loads are a certain size and you want the whole car to make it feasible. If you are shipping by marine barge, the tanks in the barges are a certain size, and so on. We have a big table for all our modes of transportation that lists the economic parcel size increments."

"OK, so what does that function actually do?"
"The setup guy, as we call him, takes all of the demands for that week, by location, and figures out how many parcels of what size and what mode of transportation based on the inventory that we have on hand. He works down the list, product by product."

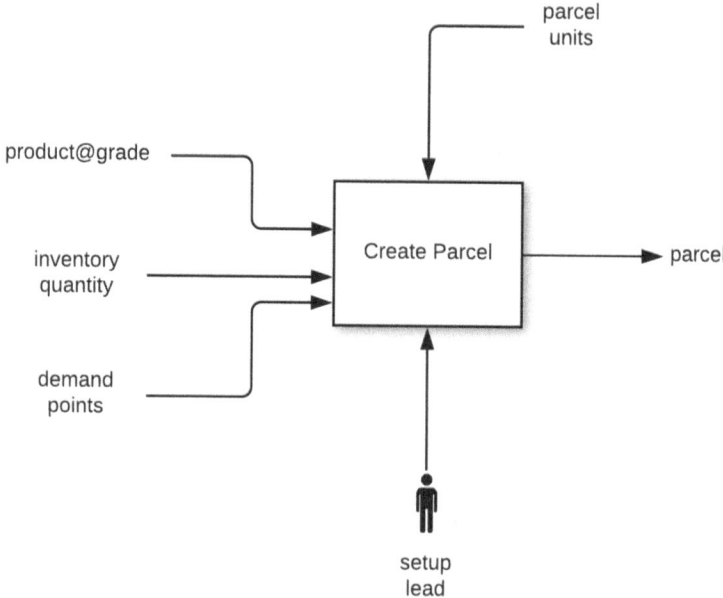

Fig. 7.2 The 'create parcel' process

"That sounds like a lot of work."

"Oh it is, and its getting worse and worse. We used to have about a dozen products each in about 3 grades. Now we have 50 products that can have as many as 10 grades. Our folks can barely keep up."

"I notice that the only resource there is the Setup Lead."

"Yes, that's right, the Setup Lead creates the parcels."

"How does he or she do all of that?"

"Oh, well Jason does that. He's been doing it for years and years and I guess he kinda has a system in his head. Also, we have a lot of repeating orders so I suppose he just does those the same way every time."

"How old is Jason?"

"He just had his 57th birthday last week."

Create Parcel appears to be one of the functions that dramatically illustrates the criteria we discussed at the opening of the chapter. The proposition on the table here is to create a Digital Twin of Jason, and all of the rules in his head to accomplish parcel creation automatically, perhaps with a human like Jason overseeing the testing and validation of the automaton.

For the practitioner, the work ahead is to decompose the single function block into its constituent sub-blocks, working down in pyramid fashion until Jason's logic is well captured. This logic can then be simulated by presenting the resulting algorithm, Jason's Digital Twin, with a set of demand points and producing a set of parcels. Did the system produce the parcels in the same way that Jason would have done? Did the system produce parcels that maximize the operating margins of the company?

In my years of working assignments just like this I often find that people like Jason do an astonishingly good job at what they do. However, like all humans there are holes in their knowledge, and rules that were once correct are now outdated under new conditions. This is why the simulation of the Digital Twin is so crucial. To realize the maximum value from our work in automation we must simultaneously mimic the human and test the human's knowledge at the same time. Improvements on the human's knowledge uncovered during the simulations are a common occurrence.

The question of *where* to build these kinds of automation solutions arises frequently. It is often impractical to shut down some actively running system in order to build and experiment with a variety of automation options. One needs a playground in which to safely play without annoying anyone who is fast at work running the business. In modern parlance this is called a *sandbox*, an idea I introduced in a previous chapter.

In a real sandbox you have a walled-off area, scattered with toys allowing children to dream and build and play without hurting anything (as long as you stay in the sandbox). For our purposes, the sandbox is a "builder's space" for practitioners. A proper sandbox should have all of the toys (tools) needed to build small functions and prototypes for experimental purposes.

A secure "private cloud" environment is set up inside an organization completely behind the corporate firewall. This provides all the advantages of the seamless cloud architecture with no extraordinary compromises with security. The private cloud is completely separate but interconnected to the Production IT environment for data sharing purposes, free of the onerous security stewardship requirements of Production IT. That does not mean that sandboxes are insecure, it simply means that they are allowed to operate in a different manner than customer-facing or critical business applications that cannot fail.

In the cloud, users can create simple functions or sophisticated models using a language that is appropriate for automation, such as those described in Chapter 6. The technical computing language is in place to serve a wide range of computing needs that go well beyond spreadsheets. The sandbox may afford the practitioner several choices in databases as well, perhaps including a graph database.

The uniform environment of the sandbox promotes code reuse. It is expected that users will take advantage of code fragments that have been created before, snapping them together in "Lego-like" fashion to enhance the speed of development and to reduce errors.

Best practice calls for organizations to create these sandboxes as a "safe" place for practitioners to play around with ideas, to experiment with the expectation that many automation ideas will fail. As pilots and prototypes built in the sandbox succeed however and reach a level of maturity, they may be spun out of the sandbox and into the Production IT environment in an orderly fashion.

Summary

This chapter represents the pinnacle of automation practice: the building of automatons from a carefully constructed diagrammatic representation of the system. As you can see the building blocks of automation are accessible to anyone with a solid understanding of the foundations of business.

Congratulations on making it through the process. You have now built your first automaton (even if it's just in your mind). Does it work? Does it work as well as a human? Better? Does it make mistakes sometimes? These are all questions that can be answered by vigorously testing the automaton, again and again. The next chapter will introduce you to the science of testing and validation, to ensure that our systems run safely and effectively and yes—exceed the performance of a human in that same role.

Bibliography

1. http://www.oliverwyman.com/our-expertise/insights/2018/may/tomorrow_s-factories-will-need-better-processes.html?utm_source=exacttarget&utm_medium=email&utm_campaign=market-pick&utm_content=2018-may_jobid_39414723.

8

Testing and Validation: Dear Rubber, Meet Road

Automated systems are unique among the mechanisms we use to conduct operations. They must not only perform to an absolute standard of performance, but they must also perform as well or better than their human counterparts. Rightly or wrongly, automation is considered a failure if it does not respond better than a human across a very broad range of conditions, including rare exceptions. "Aha!", the skeptics will say, "your automated system doesn't work because it failed to account for ___, where a human would have done that easily". This is an incredibly high bar set for us as practitioners.

We meet this considerable challenge by applying the same rigor, energy, and sophistication to testing and validation as we did during the development process. It is not unusual to spend the same person-hours testing automated systems as building them. This is especially the case for systems that have a remote chance of hurting or killing someone, or causing a large negative financial impact. The process is the same in any case, the difference is the budget of time spent on testing.

You will notice that we began the chapter with two key words, testing and validation. These are independent endeavors. Testing is ensuring a system behaves according to its design. Validation is ensuring a system meets its goals. Let's say that your goal is to bake a cake according to a recipe. The recipe outlines a number of individual steps, one of which calls for 2 eggs to be added to the mixture. Did the cook add two eggs (pass) or three eggs (fail)? That's testing, which has a clear standard to be met. Running the whole recipe with all tests passed, did we end up with a cake, or pudding? That's

© The Author(s) 2019
G. E. Danner, *The Executive's How-To Guide to Automation*,
https://doi.org/10.1007/978-3-319-99789-6_8

validation, which can sometimes have less clear guidelines. Our automated systems must be *both* correct and valid, noting that it is possible to be correct and invalid.

As a practitioner you must have a mastery of testing concepts as a means to orchestrate the proper vetting of your proposed automation, even if you are not the individual actually designing and executing the tests themselves. Since all automation consists of algorithms, and algorithms become software eventually, it is unsurprising that testing automated systems follows the traditions set down by the software community.[1] That community has in recent years moved away from a "build it completely then test when finished" sequential approach, for reasons that sound obvious to us today. Rather, the newer idea is to design tests in parallel with the development of the system, since we know early on the external outcomes of system that we expect if all of the underlying logic is working correctly. This is called test-based development, and a fundamental tool of it is called a Unit Test.

Unit testing, as the name implies, isolates a particular code fragment and applies a known test scheme to it, with clear pass/fail guidelines. So if I have a piece of code whose goal is to add two numbers together, I might set up a test as such:

Test A-1

Test part	Inputs	Expected output
1	2, 2	4
2	3, 4	7

I'm giving my code two subtests to try as part of the unit test called A-1. I run the test on my code, which focuses only on the place where two numbers are added together. The results are as follows:

Test part	Inputs	Expected output	Actual output	Test result
1	2, 2	4	4	Pass
2	3, 4	7	12	Fail

It appears that my test failed, because it failed one part of test A-1, and any one failure is enough (in this case) to deem the whole test as failed, because the adding of two numbers together should never be subjective where only a

---

[1] It should be noted that these testing and validation principles apply to non-software implementations as well: things like pseudocode, rules, and policy instructions.

portion of the test is passed. There are other more sophisticated logic functions where, in rare cases, you might declare a function to pass a test without passing every subtest, but we are keeping these concepts simple for now.

Our test failed, so what is going on? Since our test applied to a certain very specific portion of the code, we go to that section and find that instead of the "+" sign to add the two variables, we somehow added the "*" symbol for multiplying the variables. We fix that piece of the code, rerun, and have the following test result:

Test part	Inputs	Expected output	Actual output	Test result
1	2, 2	4	4	Pass
2	3, 4	7	7	Pass

The unit test is passed. Not only is it passed, the unit test is there to recall anytime I wish as the code is further developed, so in the case where the developer rewrites this section of code again and makes another mistake, that new error just introduced will be detected just the same. Most systems of size will be comprised of many Unit Tests like this one, collected into a Test Suite, to be run all at once.

There are several noteworthy points that come from this trivially simple example. First, you can see that designing the overall outline of the test doesn't pertain to how the underlying code is built. It could have been written in any computer language, using a simple, one line method or a complex algorithm with many steps. The decoupling of the test from the implementation target therefore allows anyone to notionally design tests for the system, including of course the practitioner. Best practice in fact suggests that the test designer is a different human altogether from the human who develops the software target. If not, developers who write their own tests will orient the tests to prove the code works, rather than look aggressively for potential problems.

You also may have noticed that tests have a robustness quality to them, they test the target under a wide variety of conditions to ensure completeness of the test. If we had built our test with subpart 1 only, given the inputs 2 and 2, the target would have passed the test and the error would have gone undetected, although likely caught elsewhere downstream in the code. A simple second subtest ensures that the multiply error can be detected. Often the art of testing is a function of the judgement of the test designer as to the scope and diversity of the subtests. One will rarely test each and every line of code, but rather spot check critical junction points in the chain of business logic. For each spot you can go for a high level black box aggregation of many individual steps, or granular by testing each step. As you can see, there is a range of subjective choices to be made by the test designer. Do not be intimidated by this. As a practitioner, you will usually have a feel for the kinds of tests that make sense for a system, and you will learn to devise better, deeper tests as the development ensues. To whit: test design is not a one-and-done at some specific stage in development, but rather is continuous, learning-driven, and gradually more granular detective work from the start of development to the end. My own team describes this as *co-evolution* of tests with the code.

Software developers like to think of tests in two different categories, reflecting their purpose. Testing to verify the business logic, as we have done here in this example, is one. Another is for the purpose of intentionally trying to break the system, in many cases at what are called *boundary conditions*. Say instead of inputs 2 and 2 we made another test of 2 and "banana", or 2 and the number 5 followed by 200 zeros (a really big number). In the first case does the system generate a 2, signifying that a piece of text automatically is assigned the value zero, or does it fail horribly because it did not anticipate a text input? In the second case does the math processor fall over with such a big number? We want our systems to gracefully handle odd inputs even when an output is not logical or achievable. As a corollary, we want our systems to kick out such examples (to humans) where its own computations, for whatever reason, have low confidence.

All of test testing we have discussed thus far is about getting systems to work in controlled, laboratory-like settings before they are unleashed to the real world. We try as best we can to catch all of the errors early, but sometimes *usage patterns* in the real world create errors in systems that have been shown to work perfectly under controlled test. A classic example comes to us from Amazon [1].

On a single day in December 2014, shoppers in the UK found themselves quite a load of bargains on Amazon's UK site. Items that were worth much more were selling as low as 1 pence. Consumers eagerly snapped up these once-in-a-lifetime deals for about an hour before the anomaly was detected by Amazon and selling was halted.

The next day the news media reported that a software "glitch" in pricing caused numerous 3rd party vendors to lose serious revenue. Almost all of the news stories were centered on the hapless merchants who appeared to be victims of the alleged glitch. What went wrong?

First and foremost, this was not a "glitch", which is the case where a piece of software that is supposed to work one way actually works a different way, often on a temporary basis. In fact the software worked *exactly* as it was programmed to do.

For a mass merchant, pricing is one of the most laborious functions to perform. It is no surprise that a company like Amazon that is an outspoken supporter of automation would choose to focus there. Amazon UK uses a 3rd party service called Repricer Express, a company from Ireland that specializes in retail pricing algorithms. It turns out that the algorithm constantly adjusts prices as a human would, taking into account competing prices, movement of goods, fashion trends, and the like. On this particular day the algorithm began to adjust prices of a range of *competing* products, adjusting them downward, then another, then back to the original product in a self-reinforcing loop until the prices drove themselves to the minimum rather quickly. The code performed the dynamic pricing algorithm flawlessly.

With the benefit of hindsight we have today, what should Amazon and its partner Repricer Express have done? Without leveling criticism at either firm, it stands to reason that for the test suite to be a faithful replica of the pricing algorithm, it should have included a simulation of the many Digital Twins—instances of competing product pricing agents—that were at work that day. It is very likely that the odd reinforcing loop behavior would have been observed, and the problem averted by making alterations to the underlying algorithm to make sure that it doesn't "compete with itself". This was a failure of the testing regime, not the code design, because while it is a bit much to ask code designers to anticipate every possible usage pattern, it is a reasonable expectation of the test suite to catch such obvious occurrences. In this very case the code was free of bugs, but in production the code did not perform as it was intended. We would say here that the code is tested but *invalid*, which are two independent assessments.

We see this frequently in our own work with SMEs. To an SME it is so patently obvious that you would never compete with yourself in a dynamic pricing system that they don't always think to include it in a rule for a "dumb" computer to consider. Yet gleaning a comprehensive set of information from the SME and then testing in a simulated production environment alongside the SME is precisely what validation is all about.

To help us talk about validation, I present to you Joe, The Planner. Joe works for a highly profitable swimming pool chemicals company that buys materials in bulk from manufacturers, blends and packages these materials into finished products, and distributes the products to retailers. The company maintains a product slate of 1500 individual Stock Keeping Units, or SKUs.

In his role as planner, Joe has to decide how much inventory of each of hundreds of products to carry at any given time, and orchestrate the gaggle of manufacturers to deliver the materials to satisfy those inventory levels. Carry too much inventory and the working capital costs are excessive. Carry too little and retailers will complain or move to competitors for their supply. That "just right" inventory value is something that Joe has built up in his brain in over 20 years of experience in working for the company. Joe is set to retire in six months, and his 28-year-old replacement, Jane, is on board: bright and eager to get started. However, the company has eliminated the old role called Planner, and instead hired Jane under the title of *Fulfillment Excellence Coordinator*. Her job is not to keep a lot of rules about inventory in her head, but rather to manage an automated system that will calculate the inventory levels for each product daily, then issue purchase requests to the manufacturers accordingly. Eventually Jane will manage not one pool chemical company, but several as the parent corporation is looking to acquire weaker competitors in a distressed market.

This case rather starkly illustrates the difference between automation and other technology systems. If this were simply about technology, the IT department would take months to design a specification for a new inventory management system, interview a few software vendors, select the vendor, and install and implement the system. There would be an associated change management effort to train the users and get everyone comfortable with this new style of working. Joe quietly cleans out his office and goes to the lunchroom where his retirement party is waiting.

This kind of thing happens every day, day in and day out in our organizations. However, tech-fitting inventory management is not our objective here. Our goal is to design a system that automates Joe's "algorithms" for managing inventory first, choose the technology that most easily executes that second. The pool chemical company has been successful for the very reason that highly skilled, experienced people like Joe do what they do for a living and are dedicated to the craft. In a sense Joe's brain contains intellectual property and a competitive advantage that transcends the technology of simply making the function work. Now Jane will oversee Joe's legacy in the form of multiple Joes running faster across many business units that were once companies themselves. In time Jane will develop her own intellectual prowess at managing a network of pool chemical operations, to be passed on to her successor generation.

You the practitioner patiently work with Joe over the course of weeks and months to devise the set of algorithms that set the inventory policy for the firm. To verify the system, a test is run on a sample of 100 SKUs from the master set, randomly drawn. In all 100 cases the algorithm performed just as well as Joe would have done, and in fact Joe commented that in at least 2 or 3 cases it caught trends in building demand that he would have missed.

Thus begins a month trial where Joe's Digital Twin operates in parallel with Joe himself. The trial is running very well, as the algorithms are happily humming away with calculations for inventory for each and every SKU that approximate those decisions that Joe has made.

About 2 weeks in, Joe walks into your office. There is a problem. A big problem. Joe has a worried look on his face.

One particular SKU is only purchased by one company, and then so only in small quantities twice a year. The algorithm showed stock-out (zero inventory) occurrences over long periods of time because the system prefers to keep inventory of these slow-moving items at a bare minimum to reduce working capital costs.

> We are stocking out on product HG4382. That can't happen.

> It looks like the algorithm is working perfectly.

> Well maybe so but Giant Poolco buys that product from us every year.

> OK, but only twice a year and in small quantities. Stocking that slow-moving product to never stock out would be costly, and margins would fall below the minimum the company has set for us.

> I know but we have to do this anyway, for this product and this customer alone. Giant Poolco is a retailer that owned by the founder of our company.

It would appear that the company is choosing to make an exception to its inventory policy for a particular product. That's fine, that is the company's decision to do so, and in my many years of working with organizations I've seen all sorts of exotic rules of thumb like this that come into play. To implement that exception there must be a rule that in pseudocode says something like:

> RULE GiantPoolException: Inventory of product HG4382 should never go below X units.

This rule should be run in such a way as to trump any other rule that wants to set inventory in the same way as the other 1500 items. But as practitioners we should not stop there. We should also have the system expose this rule as the anomaly that it is so that it can be re-evaluated:

RULE GiantPoolExceptionMessage: Every 6 months, send an email to Operations noting RULE GiantPoolException

What if the founder sells GiantPool one day? Unless we build self-reporting into the algorithm, this rule will silently continue to work away long after it is out of date from a business standpoint.

Adding the rule above has now rendered the system *valid*. Before the rule the system was perfectly accurate and performed flawlessly in testing. However, Joe completely forgot that Giant Pool must be treated differently, because that mode of operation was so second nature to him after 20 years of working with the company. This kind of lore will very common for you as you work as a practitioner.

It is unrealistic to expect that testing catches every error, or that every error can be anticipated at design time. Be cautious, however, because this has often been used as an excuse to short an otherwise comprehensive testing effort. On the other hand, it *is* expected that we do not repeat failures twice. Experience and history are valuable assets. A brilliant member of my staff put it this way:

> The amount of time applied to testing is inversely proportional to the experience and confidence in the task.

Our goal is to come to a Goldilocks bargain between feeble testing to let the errors flourish and become vetted by the real-world system, and a test suite that is so realistic as to be economically infeasible. Good, sound common sense thinking in our experience is almost always the best guide to setting the optimal position between the extremes.

It is very important for all organizations to build a "culture of testing" and tinkering, that was once present in our business lives, but no longer. In many cases I have witnessed a cultural resistance to testing, because testing is not building, and building is "adding value". Don't fall into that trap.

Summary

Successful automated systems got there through an equally successful testing and validation regime that paralleled their development. Systems have to be both correct and valid to accomplish the goal, and these are independent workstreams, with divergent methodologies. The fact of the matter is that, rightly or wrongly, automation has a much higher bar for accurate performance than other similar technological systems, creating a stewardship philosophy around the testing and validation process. It is one of the most critical factors of success, and should warrant your full attention as a practitioner.

Thus concludes a whirlwind tour of the nuts and bolts of building automated systems. Armed with these skills and tools, you are ready to wade into the real world of unautomated organizations, making them better/faster/cheaper … and, to invent a new word, "growier". In your journey, you will encounter both open arms and hostile resistance from companies that are not only accustomed to life without automation, but quite content with their current financial performance and see no obvious, compelling case to change. This leads us to Part III, where we will take on the topic of *institutionalizing* automation.

Bibliography

1. "Amazon Sellers Hit by Nightmare Before Christmas as Glitch Cuts Prices to 1p", *The Guardian*, December 14, 2014.

Part III

Institutionalizing Automation

Our goal here has been to create skilled practitioners of automation so that they can be unleashed within organizations to make them better, more valuable, and resilient enterprises. What about the enterprise itself? What kinds of strategic implications are brought to the fore when the company is engaged in an automation effort?

Our conversation now moves to a new level: that of *institutionalizing* automation as an organizational competency and asset. Increasingly, companies I speak with around the world are keenly interested in this. The idea of creating a Version 2.0 of the current enterprise that is much more automated is an enticing proposition on many fronts. But this does not happen by accident simply by training and energizing a few individuals as practitioners. Rather, one has to embrace automation as a business philosophy.

This part will examine automation from an enterprise-wide view. Senior executives and those that aspire to lead should take note of the outline laid out in these final chapters. Adoption of these principles will define the winners in the coming decade.

9

Automation as a Business Strategy

The Industrial Revolution began in Britain in the late 1700s. The emergence of machines for making textiles and tools at scale, combined with the steam power engines to drive these factories dramatically altered the landscape of business. Prior to the industrial revolution, people were scattered in rural communities where clothing and tools were made at home and most were employed in family-owned farms. The mechanization of basic production in factories in turn also gave rise to our modern banking system and organized, shared transportation infrastructure. Business everywhere was irreversibly changed after the industrial revolution, marking a significant inflection point in our history.

Automation will be no less impactful. It will change the way we do business today and create other unique classes of businesses as well. It also might destroy certain businesses that rely predominately on human guidance where such guidance could be effectively automated. Companies today have a stark choice: embrace and exploit higher levels of automation or risk extinction. We are already beginning to see the evidence as in the Amazon Go example discussed in Chapter 1.

The best organizations will seamlessly weave automation into their business models, making it an integral part of their strategy. But how? Let's start by establishing what strategy is, which is a challenge because there are so many definitions from a multitude of experts. I will offer my own here so that we can make progress, acknowledging that not everyone will agree with my definition.

© The Author(s) 2019
G. E. Danner, *The Executive's How-To Guide to Automation*,
https://doi.org/10.1007/978-3-319-99789-6_9

I've seen strategy up close and personal, in fact I've spent my whole career in one way or another advising companies on corporate strategy. I've seen many of them fail, and a few succeed, even spectacularly so on both fronts. My personal observations have shaped my views on strategy. The failures are, in part, my motivation in writing this book. I want to see every hardworking organization given a fighting chance to succeed.

Strategy boils down to decision-making, for a unique class of decisions that have known long-term implications to the firm. Decisions about the product line are part of strategy. Decisions to disrupt your own business are strategy. Decisions to confront and attack a competitor on price or feature spread are strategy. In every case all strategy roads lead to a definable set of decisions, many of which are made repeatedly at regular intervals.

Average firms operate in maintenance mode, considering their strategic decisions as already made for them by their assets, industry convention, or sheer legacy. There is no formal strategic process, no team accountable to strategy, no one asking the uncomfortable questions about why things are done the way they are. Don't despair if you are an average firm. They comprise most of my clients. The opportunity here is to transform the average firm to a much better one.

If you have followed the advice given in Chapter 5 (or plan to) you will have an excellent picture of the "engine" behind your organization. This the very best place to begin. If done well, you will be able to highlight the specific parts of the engine where the critical decisions are made. A firm's strategy is the sum of all of these. The essential question can now be asked: where can automation make these decisions better/faster/cheaper?

I once had an opportunity to work with a real estate firm. Their raison d'etre was owning assets (buildings) and executing long-term leases with tenants of all kinds (commercial and residential) to occupy the buildings. We worked for weeks mapping out all of the corporate functions in just the way I described in Chapter 5. The result was a pyramid of diagrams whose complexity surprised me. *What could be simpler than leasing space in a building?* I thought.

It just so happened that we were doing this work around the same time the company was preparing to report earnings to market analysts. Specifically, company leaders were expected to provide "guidance" to these analysts as to company financial performance. For a fixed asset company, that really comes down to revenue. Revenue is derived almost exclusively from lease payments from tenants, against an agreed contractual price. Back of the envelope math works just fine here: occupancy x number of tenants x average lease payment summed for every building in the portfolio equals revenue. Earnings expectations from the analysts are relentlessly increasing. Satisfy the street and you have a plentiful and cheap capital base to draw from for expansion. Displease the street, and you are drawn into a vicious cycle of poor performance that begets poorer performance.

The critical decision for the company—critical because it is the primary driver of revenue—is price. Occupancy is not a lever, but rather a consequence of price and building location and condition. The building portfolio is given (although there is a legitimate strategic decision there as well). Price is our focus and this figuratively leapt off the page to us from the diagram.

An army of people all had a hand in one way or another in the pricing decision. Given that part of the informal algorithm for price involved a percentage of capital recovery, anyone who influenced the capital expenditure in a given building also indirectly influenced the price. Operating expenses are recovered through pricing, so the scores of people performing maintenance also had a shadow effect on pricing, and so on.

Pricing decisions were frequent and routine. Every month a handful of potential tenants sought proposals for occupying space, or, existing tenants were up for renewal. Yield management systems, such as those who price airline seats and hotel rooms, were deployed to set a basic lease rate. Humans then nuanced the base to arrive at an offer to a tenant. Every month the process repeats.

Many of the leaders in the company considered pricing already automated due to the fact that the yield management system generated the base pricing numbers used as a reference. On further examination, however, it turns out that the yield management system had a vast number of "tuning parameters" that controlled the underlying mathematical model inside the system. Tune it one way, get one price. Tune it another way, get another price. The differences were material. What we have here is a machine-assisted calculation heavily influenced by its human overlords. A degree of automation to be sure, but far from the perceived "fully automated system" that the senior leaders assumed.

Having done our forensic work here, we can now ask the essential questions:

1. Does the current system ever make mistakes?
2. Is the current decision cycle time fast enough?
3. How much human labor content is involved in making the decision?
4. Is the pricing system internally consistent; does it provide the same answers time and time again controlling for the inputs?

5. Does the current system explain itself thoroughly?
6. Can the current system scale as we acquire new properties?
7. Is the current system of a form that we can monetize if we sell the firm?

The answers were predictable: Mistakes in pricing occurred all the time, cycle times were too long, too many humans were consumed, pricing was a mysterious black box with no explanations, difficult to scale, and not a component of the intellectual property. Our opportunity was to transform this pricing function into an automated, controlled, elegant process with minimal human interaction but with a high degree of human transparency. We set about creating Pricing 2.0 in the firm, an ambitious task that had never been done before in the industry, much less in the company. I lost count of how many SMEs told us that this would *never* work.

We had but one cheerleader in the beginning: the Board of Directors. Most of the board consisted of C-level executives from *other* industries who understood well the power of analytics and technology together to automate important functions. They knew that the real estate asset industry was well behind current state and was ripe for just this kind of disruptive move.

With that nudge, we were off and running. Our first order of business was to create a hypothesis, which suggested that there is an optimal price for any given unit of space, driven by a host of external factors.

Next, we developed a Digital Twin for pricing, comprised not of one human SME but actually a composite of several. None of the SMEs we interviewed were particularly shy about their views on the pricing "formula", even if it was no formula at all—"always consider this, never do that" and so on. In the end, the algorithm consisted of over 200 rules in a ladder system of sorts that paced through all of the considerations that humans balance in their mental models. The excavation of these rules was tedious, arduous work, with even the SMEs disagreeing with each other from time to time. The key to this and to any work with SMEs is to put everything you find onto a diagram. The very process of the diagrammatic exercise helps to strip away the noise and leave the core intact.

All of this simply to get to the current, human-based pricing system, albeit down on a piece of paper. Now it had to be rigorously tested and validated in the manner that we described back in Chapter 8. Sure enough, we did find that by and large, the human system was quite effective, yet, there were a few small, undetected holes in their own logic that we found by running the algorithm against thousands of sample pricing problems and observing its results. Today I am happy to say that this company is a leader in its industry and never mentions publicly its "secret sauce" tucked neatly away under the hood of the corporation, humming away day in and day out.

What if the board had not been so keen? Is there a risk to *not* automating firms?

Just ask UPS.

Notorious for miserly spending on automation in its distribution hubs, it lost significant ground to FedEx and even Amazon's own distribution network just as online commerce was growing at double-digit rates year on year, each of the latter being well over 90% automated. As of this writing, UPS is playing catch-up, having to spend $20 Billion to retrofit aging hubs just to keep up with the demand. The stock price took a tremendous blow in January 2018 when the capital plan was announced. It will be years before they catch up to their more automated rivals. Watch the news media for hiccups along the way, especially during peak holiday periods [1].

UPS is filled with intelligent, skilled personnel. By all accounts they are a well-managed firm. What went wrong?

Without knowing the company (my firm has never done business with UPS or its suppliers), I believe that I can guess exactly what happened based on my own experience in seeing this pattern over and over with other firms:

1. **Fire fighting**. Like many companies, 99% of the working time is spent fighting the latest fire, not thinking about longer-term planning and improvement. I find this particularly prevalent in publicly traded companies where the quarterly focus trumps all other goals.
2. **Worse before better**. Automation comes at a price: upfront investment, some level of disruption, and resource diversion. In other words the system actually pays a mandatory short-term penalty in order to gain the long-term benefit. Many companies have zero tolerance for any kind of penalty and therefore never engage in good medicine.
3. **Where's the guiding light?** I can almost guarantee that there was no metric sitting on an executive dashboard somewhere inside of UPS screaming out "Automate or Die!". Rather, those same dashboards probably patted them on the back because package volume this month was 2% better than package volume last month. Backward-looking indicators alone rarely tell the whole story, and almost never present a strategic picture.
4. **Uncomfortable questions**. Who within UPS is asking uncomfortable questions like: Why do we do it this way? Why are we happy with that level of performance? Do other industries do something similar to us? Do they do it better? My experience is that such people, as vital as their voice is, rarely last long inside of large companies.

All of these factors conspired to keep a good sound company, well, just good, not *exceptional*. Automation is about taking good quality ingredients like talent and unique assets and making them exceptional through the transformative power of science in operations. In hindsight UPS should have taken its time and woven in a controlled, methodical process for automating distribution in surgical doses, well under the radar of the spending hawks among institutional investors. Is your organization like UPS?

Speaking of competition, could competitive moves be automated? We've already described the pricing example from the real estate firm, which was in part a competitive response, as it considered the tenant's other alternatives. But what about systems that *automate the process of competing*?

To answer the question, we should look to the science behind the strategy. Game Theory was created by British logicians in the late 1920s as a systematic way to go head-to-head in naval warfare. If my enemy makes this move, I should make that move. The field later developed into a first-class branch of mathematics from the work of John von Neumann, perhaps one of the most prolific Mathematicians of the twentieth century. Since then, Game Theory has been credited with the generation of keen insights into everything from OPEC's cartel cohesion to auction behavior for pieces of the public telecommunications spectrum. It truly is the science of strategy, injecting math into a discipline dominated by subjective art for so many years.

A classic example from competitive Game Theory is known as the Prisoner's Dilemma. It is a toy example used to illustrate similar dynamics that arise frequently in business and politics. In Prisoner's Dilemma, two criminals are caught and arrested, then separated so that they cannot communicate with each other. Each is offered a deal:

1. If you implicate your partner and he implicates you, each of you will serve two years in prison.
2. If your partner implicates you but you remain silent, your partner will be set free and you will serve three years in prison. Same deal for your partner in reverse.
3. If you both remain silent, you will each serve one year in prison.

Let's look at the payoff matrix for this "deal" in Fig. 9.1. Interesting. You can see that for either man, betrayal is the better strategy no matter what the other does (a–1 beats a–3 and a 0 beats a–2). Yet if both men had been able

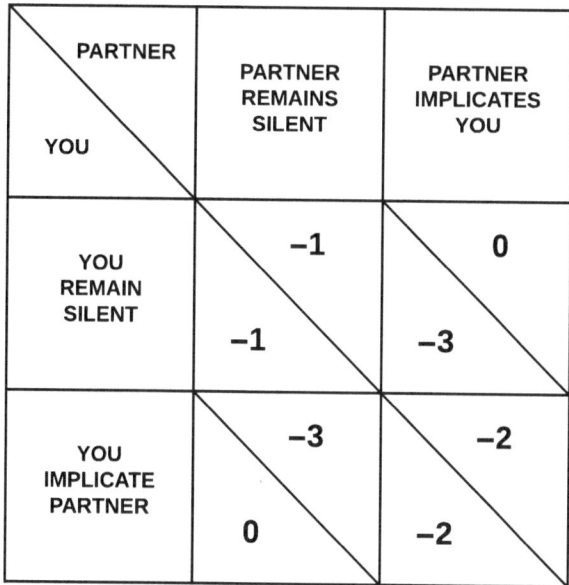

PARTNER / YOU	PARTNER REMAINS SILENT	PARTNER IMPLICATES YOU
YOU REMAIN SILENT	−1 / −1	0 / −3
YOU IMPLICATE PARTNER	−3 / 0	−2 / −2

Fig. 9.1 Prisoner's dilemma payoff matrix

to communicate with each other and see that their collective best strategy is to both remain silent, they are better off!

Prisoner's Dilemma is a classic examination of best decision-making under competitive pressure, and the counterintuitive results that unfold. Robert Axelrod, a researcher at the University of Michigan studied *repeated* Prisoner's Dilemma—the game repeated round after round with the same partners. In fact, he held the first tournament in 1980 and challenged game theorists around the world to come up with the most effective strategy algorithm that would win across many rounds of the game. The winning strategy? A simple tit-for-tat. Cooperate with your opponent on the first move, and then do whatever your opponent has done on the previous move. As simple as that.

The point of my story here is not to teach you Game Theory. Rather, my point is that competition, like any corporate function, can be distilled down to a set of algorithms, even simple but powerful algorithms like tit-for-tat. Once in algorithm form, the strategy can be automated. Imagine a cluster of software-based bots that constantly check your operations against the market offerings and makes ever-so-slight adjustments in real time. The algorithms would be smart enough to anticipate your competition's next move and counter for that as well. Welcome to automated competition.

The investor cohort is one that is keenly interested in automation, and it doesn't take a stretch of the imagination to figure out why. Buying an unautomated but asset—and talent-heavy company at a book value then automating that company generates high value for the eventual sale. It's rapidly becoming a primary part of the investor's playbook. Automation reduces cost per produced unit, and cost leadership in the market is *always* a winning strategy, no matter which way the markets move.

In my nearly four decades of observing companies, certain enduring structures tend to dominate, and not in a good way. I'm talking about organizations that are constantly guided by brute force by an army of humans to seek out a dollar here and there. Unfortunately, most of the time they cover their costs and earn a profit, giving the mediocre the right to do it all over again the next year. What remains invisible in these firms are the hidden vulnerabilities due to a vitamin deficiency of resilience. Pick up any business journal on any given day and you can read all about it. The phrases "unravel" and "vicious cycle" are used repeatedly, both describing a lack of resilience.

What I find most interesting is that in recent years many of the stories start with an uptick or surge in business and go on to show how this great fortune actually turned into a weapon that ultimately pulled the firm under. How can this be?

The fact of the matter is that resilience under surge and other disruptions is a yet-to-be-discovered art of leadership. Very few firms have mastered this, which is why we see stories in the news every day about the public struggles of organizations in dealing with the surge. In fact, nearly every Christmas season that I can remember there have been reports of this company or that buckling under a burst of new orders.

The ultimate goal of incorporating automation into business strategy is to create organizations that are *autocatalytic*: give them a nudge in the right direction and they continue to operate in the correct direction until they are nudged, ever so slightly, again. Highly intelligent feedback loops keep the firm operating at near-peak performance between nudges. At the same time, these systems are also built to be responsive to external disruptions like surges in demand, because they have been thoughtfully pre-programmed, via other special sets of loops, to respond just in the right way.

Autocatalytic firms use a profoundly different process to design their strategies, relying much more on carefully crafted algorithms than on human experience and knowledge alone. In the coming wave of automation, these will be the firms that survive.

Summary

I wish I could tell you that as a business person you do not have to worry about automation. You are in a fine company, that is doing well…just relax and rest on the market prowess that your firm has been exploiting for many years.

But I can't.

My years of observations have led me to conclude that all businesses have a blunt choice—automate or suffer. Yet the news here is not all bad. Businesses can adopt automation as a business strategy and incorporate its power into the firm's operations using the guidance laid out in these pages. The best place to start are those key leverage points in the organization that have the most impact on the final financial performance of the firm.

For a practitioner, the goal is to move from one leverage point to another and another. In short order, you will find yourself with an entire enterprise using pockets of automation to perform its most important functions. When this happens, you will have graduated to an entirely new level and will have to consider the organization as a whole. The next chapter leads us to considering an organization in an end-to-end way, and the opportunities for enhancing automation at critical mass.

Bibliography

1. "UPS's $20 Billion Problem: Operations Stuck in the 20th Century", *The Wall Street Journal*, June 15, 2018.

10

Automating an Entire Corporation: Baby Steps on a Long Road

In 2018 as this book is being written, there is a hopeful view that manufacturing will return to places like the United Kingdom and the United States where it once was the bedrock of these economies. The evidence suggests that this is a very real possibility, with the economy, tax law, and regulatory policy all conspiring to bring the process of making goods from far-flung nations back "home" [1]. This is not a political statement but a factual observation of trends.

When manufacturing does come back to these countries, it will not be the same as it was 30 years before, simply transplanted. Recall the story of Joe the Planner from Chapter 8? Now Jane is in charge, and it is a very different system, albeit the goal to produce pool chemicals profitably remains the same. Jane's landscape, characterized by smaller numbers of human coordinators working with automated systems at their behest—is the likely format for manufacturing returning to their home countries. Levels of automation will be reflected outside of manufacturing to all of the attendant industries—services, transportation, energy, communications. The expectation everywhere will be that automation is not an optional luxury but a compulsory prerequisite for staying alive in the market. My job here is to get you ready for that future.

To do that, we need to get our organizations ready. That is what this chapter is about.

The landscape in the corporate world is daunting. Most organizations these days fight fires—that's all they do every day. Often, I see resistance to automation because it "gets in the way" of fighting the latest fire. Don't let

© The Author(s) 2019
G. E. Danner, *The Executive's How-To Guide to Automation*,
https://doi.org/10.1007/978-3-319-99789-6_10

this mentality dominate. There is serious disruptive work to be done. What is even worse is the firefighting CEO, and I have met a few of those along the way. I mentioned firefighting in a previous chapter, but this chapter will show how to counter this negative gravity.

The hidden disease inside of nearly every corporation in the world is this: Institutional Knowledge. Not that IK is a bad thing, in fact it is the very core of what great firms acquire to run so well every day. The problem is the packaging: human brains. I often ask my clients to do an experiment: take all of the major cognitive tasks that the company sees as its competitive advantage and make a list of them. One pharmaceutical client of mine, for example, listed the fact that they have mastered the "just right" balance between in-house drug development and biotechnology partnerships to buy down risk.

Then ask the essential question: go down the list and mark how many of these things that are done my human experts, and how many are carefully documented and codified in a system. The vast majority are frequently in the human brain category. If most of your IK is in human brains, your company is at risk, plain and simple. Humans can quit, retire, die, become disabled, be hired away by competitors—all resulting in an immediate and permanent loss of valuable IK. Humans are good at many things, but we should not use them as erstwhile hard drives to store all of the critical information that makes the company run. Automation forces us to do what we should be doing already, thoughtfully and methodically extracting IK from its organic source in the human brain and putting all out in abstract form on paper for eventual codification in an automaton.

Many companies struggle with the very first step of automation, creative ideation about what to automate and how. This is the root cause behind many organizations, even well-managed ones, failing to get onto a program of continuous automation in spite of the desire to do so. Can ideation be learned, even when there is not a creative bone in any single body in the entire company?

There are several ways to throw a catalyst into the mix to get automation ideas moving. I find Pyramid Thinking [2] to be an excellent way to guide a group of bright people through a thought process. As the name suggests it starts with a plainly stated end game vision then connects the supporting arguments in layer after layer until you reach a logical conclusion around each thread. Done this way it is not difficult to get an audience of experts weighing in. A second approach that I find equally successful uses a technique known as Value Mapping. Here, a facilitator asks leading questions about the future state of an organization, while the audience participants individually write short phrase ideas on sticky notes. The notes are placed on a board for all to see, and are then categorized, de-duplicated, and connected to each other with flow lines drawn by the facilitator. One of my favorite sidebar exercises for groups is to find case studies of organizations *outside of their own industry* that exhibit characteristics that we wish to adopt. Cross-industry learning is an extremely valuable activity that is practiced far too seldom in industries where it is often assumed that their industry is so special and unique that it cannot possibly learn something valuable from another. In my 35-year career of studying organizations around the world in every conceivable industry, I can assure you that nuggets of wisdom and ingenious solutions exist in every corner if we simply take the time to look with a critical eye.

Try this thought experiment: How would Steve Jobs or Elon Musk or Jeff Bezos or Richard Branson run our company if they could? The results of that experiment will surprise you in a number of helpful ways.

As I mentioned very early in this book, *Systems Thinking* is fundamental to the kind of end-to-end thought process that is required for serious work in automation. The most famous exercise for teaching Systems Thinking to a broad group of individuals is something called The Beer Game.

The Beer Game is a board game consisting of competing teams of 8 players, further divided into 4 pairs each affixed one of 4 positions in the value chain of a fictitious Beer industry: the Brewery, Distributors who buy from The Brewery, Wholesalers who buy from the Distributors, and Retailers who buy from the Wholesalers. At the very end of the chain are customers, whose demand is represented by a deck of cards. The objective of the game is to keep up with demand by buying beer but doing so in a way that minimizes inventory. The team with the lowest overall score, as measured by inventories and backlogs, wins.

A 4-part, linear supply chain. What could be simpler than that?

The subtleties in the game lead to very surprising outcomes in spite of the simplicity. More importantly, it teaches the impact of delays, reinforcing loops, information asymmetry, complexity, decision-making under pressure, and the role of systems—in other words, many of the fundamentals that are used in the automation world. For those companies that want to orient their thinking around end-to-end views of the "system" that is the organization, you cannot find a more effective exercise than the Beer Game. Anyone, from the mail clerk to the CEO, can (and should) play. I find that companies that play The Beer Game come away with a much deeper appreciation for the connectedness of organizations which in turn provides an intellectual tailwind for discussions of automation.

Corporate Hackathons are growing in popularity, and might also be a useful catalyst for automation. Originally designed to engage in overall innovation in the business, hackathons can be steered into an automation theme rather easily. The best teams consist of a diverse mix of technologists and business people, sequestered in a room, supplied with ample food and drink, and report their invention at the end of the time period. The invention in this case does not have to be any kind of working prototype but rather could be a well-designed automaton on paper with all of the ingredients gathered together to concretely express the idea and its value proposition.

All day/all night hackathons writing code are the norm in the technology world, but I find the all-night aspect to be a bit overkill and unnecessary. A reasonable alternative might be a 12-hour stint conducted on a Friday so that the participants can get plenty of rest the next day, perhaps with an additional half-day break on Monday morning.

The results from a "create-anything-you-want" hackathon are rarely successful. The preparation should narrow the focus to a few key themes with some suggestions for associated algorithms that would be useful to the company. Again the goal is to get enough of the algorithm working after a fashion to explain what it does and how it works. The most promising of these can then be developed further by practitioners into full-fledged prototypes.

Even for the creatively challenged, these techniques can draw out the ideas in a logical, clear manner, and form the Genesis for an automation program. Since automation is a continuous, adaptive endeavor, getting started in a less-than-perfect way is far more important than spending inordinate time trying to get everything just right in the beginning. I've seen many automation efforts go nowhere while waiting for the perfect entry point.

You will likely come away from an ideation exercise with not one but many equally valuable automation ideas, alongside other ideas that have been "floating" in the halls of the organization already for years waiting to be implemented. Best practice in this area is to put all of the ideas together onto a roadmap that shows the progression of the firm from current state through to several future states in progression. The roadmap is a highly structured explanatory picture of work spread out over time. Here are some guidelines for crafting an effective automation roadmap:

1. The horizontal axis is always time. For most corporations, five years is a good time window for an enterprise-wide automation program. Each year 0–5 should be shown as a column.

2. The rows comprise the "how" aspect of the program. This might include the following categories: projects initiated, level of investment, expected outcome, technology foundation, target processes, people, and culture.

3. The cells define as specific as possible the action plans or the effect of the program or both. A good roadmap is one where the cell on the left lays down capability or infrastructure that is used as leverage in the cell just to its right.

4. Best practice is to create a roadmap using a cloud-based technical drawing tool like LucidChart. In this manner, everyone with a browser can see the roadmap, yet it is easy to update as the program unfolds. It is also easy to add notes and comments in layers that can be turned on and off.

5. There is a tendency in some roadmaps I've seen to make them overly wordy. Engineers especially (I'm one so I know) tend to want to be ultra-precise in their specifications of the cells, and therefore over-explain what is easily accomplished in a simple phrase. Again tools like LucidChart can be your friend—a simple phrase on the roadmap can link to a much more comprehensive "wiki" document that explains the means of the phrase in detail.

6. Once done, print it poster size and put a copy in the lunchroom, on the bathroom stalls, next to the copy machine, by the coffee pot. A roadmap doesn't do much good if the copies are squirreled away on some hard drive somewhere—socialize it across the organization.

Next comes a consideration of the talent needed to perform all of the tasks and projects laid out in the roadmap. Corporations are justifiably wary of hiring outside consultants for anything, much less for automation. On the other hand, the talent found inside does not likely have the requisite skills to become practitioners. Hiring for the specific purpose of creating an automation team is slow, expensive, and painful.

My recommendation is something I refer to as the Hybrid Model (see Fig. 10.1). It begins at the outset with heavy emphasis on the use of experienced outside automation practitioners, but with the important stipulation that internal resources run alongside the practitioner observing and absorbing the process. I will acknowledge that it is the rare practitioner who simultaneously does his or her work and is also transparent for a group of untrained observers to learn the practice. Yet, this stipulation is essential for the Hybrid Model to work. The end effect is a gradual tapering of the external experts in favor of the internal resources, usually accomplished over a handful of projects.

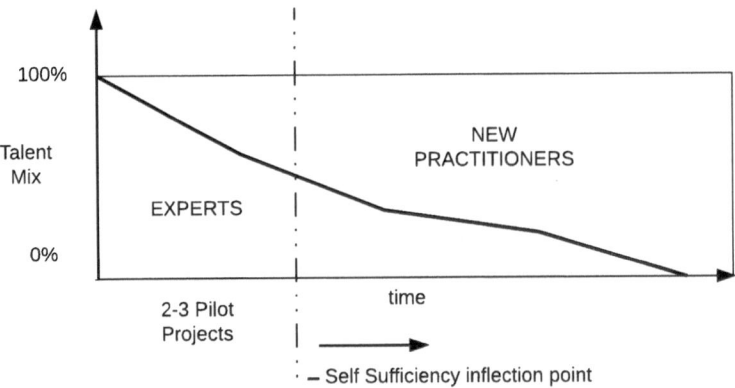

Fig. 10.1 The hybrid model

Even if you are years away from maturity in automation, you can begin preparing an internal talent base. I refer to them as "sleeper cells". Most firms cannot afford to take their best people, divorce them from their day jobs to place them in this automation specialty whose importance and level of intensity is months or years away. However, you can take individuals who show promise in understanding the business, technology, have diverse interests, and exhibit an elevated ability to learn new things quickly. Take those individuals and train them in automation concepts. Provide them with research materials (like this book!) to study in their own time. But leave them in their roles running the company's operations. In this manner you have created a sleeper cell of talent that can be activated on a moment's notice. The speed of reaction may be important as the automation wave will come upon us faster than organizations normally react.

Inevitably, someone in the enterprise will ask: where do we get the data? Data (customers, transactions, assets, suppliers, products, employees, locations, processes, inventory) is certainly the fuel for any serious work in analytics, and automation is no exception. "Our data is terrible" is something I hear a great deal from my clients. "With the state of our data there is no way we can advance into something as sophisticated as automation".

Most of the time this obsession about data quality is unwarranted. Perfect accuracy and comprehensive coverage are not necessary for most kinds of automatons we are describing here. The view that even the slightest inaccuracy or void in the data throws the entire algorithm into the "garbage" category is simply not true. Even so, data is important.

The good news is that many of the companies that I encounter of any size have invested in the last few years into an Enterprise Resource Planning (ERP) software backbone, such as SAP or Oracle (and a myriad of other applications in the SMB market). If your organization has done this, the odds are that you have, for other reasons, also applied the proper health and hygiene to your data to bring it up to a level of quality that is firmly acceptable for automation purposes. A few other firms have taken this a step further by starting corporate-wide efforts under the banner of *Digital Transformation*. Done well, Digital Transformation can be a powerful enabler of a follow-on automation stage.

Keep well in mind that automation for an enterprise is not a technology play, it is an *economic* move. As such, automation is subject to the same financial return expectations that any other economic entity in the firm—assets or new product lines for example—must achieve. The best firms have an enlightened CFO who subscribes to this idea and actively engages in the automation program. To help that CFO, practitioners should work jointly to build ROI models of the automaton candidates under consideration. It stands to reason that such models can be templatized in a spreadsheet to shorten the time in building each business case.

While we are up here in the C-Suite, let's talk about the automation of that important area. No, I am not talking about a CEO "bot" that runs the company (yet), but rather a series of automatons that work very well in this domain, all falling under a framework known as IFTTT (If This Then That).

IFTTT is both a product and a concept. The product IFTTT is a popular consumer app that helps to automate common tasks via a simple scripting method. An example might be:

> When an email arrives from Bob and the subject line contains the text 'Softball Tonight', update my Google Calendar to set the game time.

IFTTT is a consumer platform for a whole host of other apps and devices to sense anything from the arrival of the pizza delivery person to the fuel in your gas tank and then perform messaging and other tasks as actions triggered from these sensors. IFTTT has become an extremely popular integration utility for thousands of 3rd party devices and applications, wiring them all together with a simple scripting language that non-programmers can use easily.

The IFTTT concept is the whole broad idea of scripting as glue between a sensed event and appropriate action. Taken as a concept, IFTTT is now portable enough to apply in industrial settings for the purposes of automation. So imagine this IFTTT script:

> Whenever my competitor announces a > 3% change in their pricing, re-run my revenue model for next quarter and distribute the output to my 5 Vice Presidents

This handy little "bot" runs every day looking for that price change, trapping the event, then inducing action by sharing the implications to the people who matter to the consequences. It automates what normally would only happen by chance, yet this might represent an extremely important finding for a corporation.

The business conditions that gave rise to the use of mechanisms like IFTTT came from the complete about-face of data. A decade ago it was hard to move analytics along because we often did not have enough data to make it work. Today, it's an embarrassment of riches, with such a volume of data coming at us every day that we have little chance of managing it well without help. And consider that the core problem today isn't just the volume of data but rather the simple fact that 99.9999% of this data is not actionable. As a business person I care much more about the 0.0001% that *is* actionable. The IFTTT concept delivers on that ideal.

Summary

Expanding our reach for automation across an entire company is where the real value from automation is realized. Getting started is the challenging part, but thankfully there are tools out there to help ignite the creative side in us that is waiting to come out. The challenges of course don't end there, as most organizations seem to have an immune response to anything that smells like change (God forbid!).

Ultimately, enterprises are comprised of humans expending their energies to get corporate functions done, day in and day out. This begs the question: What happens to the humans as we move through our roadmaps toward a much more automated version of the company than exists today? Where do they go? How are their job roles shifted? The next chapter examines these very questions: what to do with the humans.

Bibliography

1. "Reshoring Is on the Rise: What It Means for the Trade Debate", *Industry Week*, April 13, 2018.
2. http://www.barbaraminto.com/.

11

Where Do We Put the Humans?

The idea that humans will have nothing to do in a highly automated world is the height of nonsense. It also completely ignores our history. We have arguably had machines in one shape or another for over 6000 years, and there has *never* been a moment where large groups of humans had no productive channel for their energies. There is always a vocational role for humans and there always will be, but that role will indeed change over time. Short-term displacement of certain skills will occur, as the inevitable price of progress.

The future wave is no exception to this dynamic, and the way that automation is shaping up for our near future is providing us clues as to how humans will work with, for, and on machines to move society every day. My own view on this is optimistic. You will note that in Chapter 7 we introduced the idea of creating more rewarding and valuable roles for humans through automation. On net, automation will make life safer and easier for human workers, while at the same time lowering costs and increasing variety for human consumers.

Let us answer the question posed by the chapter's title: Where do we put the humans? The best way to answer it is to take on the mirror image to the question: Where do we *not* put the humans [1]? What are some corporate functions that are clearly the better domain for machines, leaving the humans to do *everything else*? Here is my list:

© The Author(s) 2019
G. E. Danner, *The Executive's How-To Guide to Automation*,
https://doi.org/10.1007/978-3-319-99789-6_11

1. **Many calculations** have to be performed quickly to reach an optimal decision. For example, a forecast of demand next month for each of 1000 individual products, or processing a camera image to count the number of faces.
2. The **tacit knowledge** and logic of how to perform the function can be expressed on one (large) sheet of paper or written on one whiteboard. Pull out the instruction sheet of your favorite board game and you'll get the idea.
3. **Scale is a must**. What you performed last year, now do 3× as much this year. Next year, you must do 3× of this year, and so on. Online retailers have embraced automation to manage surges during holiday periods [2].
4. **Prediction**. Machines have no optimism biases for prediction and typically will beat humans in predicting future events (like the weather).
5. **Signal processing**. Taking in more than 2–3 sensor readings and deciphering the meaning of it.
6. **Data transformation**. Many jobs, believe it or not, involve taking this report and retyping into that spreadsheet, then taking that calculation and punching it into another web form, etc. This is simple data transformation but hidden behind some fancy job title that overstates its underlying sophistication.
7. **Search**. To accomplish a function, I take a search imperative like "Find the ___ of the ___" and look up the information, often contained in disparate systems. Automation is far faster than humans in these activities.
8. **Execution of feedback loops**. Imagine in your primitive house you have two buttons: heat and cold. It's too hot, so you go press the cold button. After a while it's too cold, so you then turn on the heat button. Every day you constantly go back and forth with heat and cold fussing over the temperature. Silly, right? Today we have thermostats for that purpose that do a good job of maintaining temperature at just the right level. Yet, I have observed many an organization that works precisely this way. They have feedback loops and use an army of humans to press the heat and cold buttons. Because the loops are subtle, it is often hard to see it in place, but once recognized this becomes a prime candidate for automation to take over.

I am sure that with some quality thinking you can devise additions to my list presented here. The point is that we have a very large number of humans inappropriately tasked with rote, mechanical, repeatable functions.

I made a recommendation once to a company in the natural gas transportation industry. "Create a job", I said, "where the point of the job is to fire yourself". You could hear crickets chirping at the other end of the conference table.

What I meant by that was a job where there is a substantial reward for automating your way out of it. If I have been in charge of, say, regulatory compliance for a natural gas pipeline, a big part of my job is to furnish reports and data to various government agencies. What if I received a year's salary bonus for automating that reporting function, followed by a promotion? The kinds of arrangements are called *gainsharing contracts*, a phrase we introduced back in Chapter 2. They are aptly named as working to share the gain between employer and employee, versus the more traditional Principal-Agent problem where employee incentives are largely misaligned with company goals.

Gainsharing, on the other hand, aligns individual incentives perfectly with the company's goals. Reluctantly the company tried this in limited form and the results were surprisingly positive, especially for the younger workers in the firm who seemed eager to embrace automation as a means to move on to something bigger and better within the company. We asked them a provocative question: "If you could endow a human with infinite speed and data omnipotence, what could you do with that superpower? How would your job change?" To our surprise, they jumped at the chance to form an answer and came up with a long list of creative ideas to pursue. One young manager remarked, "I never really knew why we did it the old way in the first place, it seemed so outdated and inefficient. Finally, I've been given time and a reward for looking at doing it differently. That makes it fun!"

Younger people coming into the workforce is a fascinating study in human behavior. Their worldview is shaped predominately by their social and personal lives, which are highly automated. They carry these expectations into their stodgy employers and are horrified as to the primitive ways of working. As these workers mature and take on more responsibility you will see an even stronger tailwind of automation via their strong support of it.

Younger workers are also accustomed to the idea that there is no need to avoid judgment, reasoning, and other seemingly human-like qualities among the functions we decide to automate. One of the very first recreational uses of computer technology starting in the 1960s was getting computers to play games, where good judgment is the linchpin to winning the game. All the researchers did was to boil the concept of judgment down into a set of decision rules, combine them with feedback loops, and voila—you have a machine that can easily execute judgment in similar ways to humans, but much faster and with a wider stream of incoming and historical data.

Artificial Intelligence adds even more power and fidelity to the expert judgment equation. At first glance the introduction of a brand new product into a portfolio of products might seem like a purely human endeavor, and in fact today in almost every case some human Product Manager is deciding that a new lip gloss with thus-and-such features must be offered. But what if we had a system that constantly scans our product portfolio against competitors against buyer preferences and finds market voids where a "suggested" new lip gloss might fit? While you have not entirely automated product management, you have in fact leveraged the human with automation for a vital portion of the product management function.

In this case the human product manager is the master, and the algorithm that performs the suggestion is the servant. That's a normal and comfortable role for humans to play these days and in fact this is how automation will evolve at first—from human only to the machine-leveraged human, again with the human still in the driver's seat. We will see most humans actually eager to adopt this kind of automation as it raises their own performance to a much higher level. The woodcutter with a chain saw is a much happier fellow than the woodcutter with an axe.

However, automation will not stop there.

It will continue to march on and will flip the master/servant relationship. The lip gloss product algorithm will be promoted to make all of the decisions regarding the product portfolio, if it continues to perform well. The old human job role will not be filled when the human leaves the firm. Rather, a skilled human technician will simply make sure the algorithm is running properly all of the time. In this sense, the human is the servant to the machine that is the master that must be cared for every day. This is a far more unsettling picture for humans. Walk into any Amazon Fulfillment Center and you will see that the few humans that are there are in place to make sure the machines are running, not in the main line of package delivery.

The revolution of the master/servant relationship is neither good nor bad but is in fact an unavoidable feature of the automated world. How to transition human workers into this new world will be one of two formidable challenges presented to senior leaders of the era. The other is decisions regarding the number of workers necessary to keep operations going.

The assumption that the motivation of senior executives is to massively fire as many human workers as possible through automation is simply wrong. In writing this book I had the opportunity to converse with hundreds of CxOs all over the world. Not a single one mentioned mass firings as part of the plan, even off the record. Rather the overwhelming stimulus for automation lies in the relentless quest for productivity. Senior leaders wanted to take their existing worker base and produce 2×, 3×, or even 10× the output of today. Paradoxically that makes the human workers on board *more* important, not less, as we have mentioned numerous times before. If there is an overall theme to the new roles for humans it is this: humans will be less tangled up with the direct input-to-output function whatever that might be, and more aligned with a coordinator's role—orchestrating a suite of algorithms that may span business units or even companies. This in turn raises a set of skills that today are not as prominent as they will be: troubleshooting complex systems and end-to-end systems thinking.

For automation practitioners, the immediate challenge is to re-craft job roles from the unautomated "as is" system to the new, "to be" automated system, as we saw with Joe and Jane in Chapter 8.

Human workers should be transitioning to *exception-based thinking*— designing with problems (failure) in mind: original thought, first principles. You might be surprised how many people are involved in jobs today that are incredibly simple-minded. I suspect that most people in these job roles do not find them rewarding but rather just remain in them to pay the bills.

Exception-based thinking is creatively exploring what could go wrong, and then just as creatively designing intelligent countermeasures as an antidote. I once worked with an East Coast US oil refiner and trading company on a series of trading models. Several people noticed that the models were thrown out the window whenever a severe disruption occurred, like a hurricane or a fire in a compressor station at a major pipeline (these are actual examples). "Its every man for himself", the lead trader told me, "we just do whatever we can to survive with a patchwork of trades it until normalcy returns to the markets in a few days". "How often do these happen?", I asked. "About 3 times a year we get a major disruption."

What ensued after that was a rather lively discussion about refinery profitability. Most of the time refiners earn a surprisingly small spread between the cost of the chemical reactions of the refining process and the terminal price of the gasoline, diesel, and jet fuel products that are created. The market is stingy—even the cleverest refiners earn very little positive difference from the average. During disruptions, however, all of the market rules are involuntarily suspended by nature, and the spreads can widen to an extreme degree. Quite by accident some refiners make a full quarter's worth of profit goal in that one disruption. This begged a question: Could refiners actually develop a strategy to specifically *exploit* disruptions? "Yes", the trader replied, "but it's really complicated, and no disruption is exactly the same".

Over the next week, the trader and I locked ourselves in a conference room fitted with two very large whiteboards. We drew, erased, and drew again as we put together an algorithm that would predict, sense, and then respond to a myriad of possible disruptions by executing trades that would flourish in the ensuing market melee. The skeptical trader was right, it was complicated in the end, but less infinite than his mental model was suggesting. The company still uses a form of this algorithm today and has reported great success in boosting profitability.

Exception-based thinking means playing around at the margins of company operations, shining a light on corners of that microeconomy that are often deemed backwaters. Creative intellectual exploration is precisely the kind of thing that is hard to automate, and is therefore a suitable place for humans to put their energies. Ironically, people often tell me that they don't have time for this kind of thinking because they are too busy doing the repeatable mechanical stuff.

As a practitioner you should be aware that a lot of references these days to "AI" is really about automation. Not a day goes by that I don't see an article land in my inbox suggesting that "AI can do this, AI can do that, AI will help us ___" and so on. When you dig a little deeper you find that the author is really talking about machines (software) that are intelligent enough to take over some or all of a previously human-dominated function. AI is simply one of many mechanisms to choose from to get there. The next time you see an article espousing the wonderful benefits of AI, simply substitute the word automation and you will find that it fits just as well. I point this out because there is a lot of good writing about automation out there in the technology media, although much of it does not carry the automation label so prominently.

I'm often asked in the Q&A session of my public speeches if companies ought to have an Automation Department/Group/Team. My answer has evolved over the years. Early on, my response was patently "no". Automation, I argued, is a creative, thinking endeavor that should not be bureaucratized by the organization. Perhaps I had read too many Dilbert cartoons to be cynical about Automation as a formal Department, which by definition means there is a Head of the Department who knows little to nothing about automation. I also feared that an Automation Department would send a message that automation is owned somehow by a select few and practitioners and enthusiasts for the craft could not live outside of the club.

Such is the view of a snob and a purist as I was back then. Since then however, I have seen the downside of the antithesis of an Automation

Department, let's call it organic automation, arising from the passionate efforts of a few committed people. Yes, they are creative, and yes, they are boundless, but organizationally they are powerless. Budgets, special software, permission to do something different—all of these are roadblocks that even the most passionate people cannot overcome. Sadly, the organization ends up losing most of these talented and energetic folks.

So maybe there is something to the idea of creating an official "home" for automation that has a formal endorsement from the senior leaders? Is there an optimal corporate structure for institutionalized automation?

Centers of Excellence (CoE) emerged somewhere in the 1990s as a configuration of the matrix organization. These come in every shape and size, so let me describe a CoE tuned to the unique properties of automation here.

In a CoE, there is a technical leader, but that person doesn't "own" automation in any way, shape, or form. He/she is simply an internal evangelist for the craft and floats across the organization seeking and identifying demand for automation that is not always obvious. Moreover, for anyone in the CoE there is an understanding going in that all members rotate in and out of the CoE between active, full-time participants, and free agents who have other day jobs but could be tapped quickly as critical needs arise.

The CoE does, however, have an operating budget and a collection of assets—mainly a cloud-based development environment, which we will describe in the next chapter. CoE uses its funds to pay for pilot projects, attendance at conferences, and internal brown-bag talks across the company. The CoE maintains an internal wiki with all of the meta-data about the team: projects, past and on-going, members and their contact information, a roster of skill sets. What I have described is a CoE that can work, providing an optimal structure in between organic and formalized.

Automation today is on a steady march, and humans openly worry about its consequences. In my public speeches I will have professionals like attorneys and engineers and journalists come up to me afterward expressing concern: is my job going to be automated one day? My not-so-happy answer is *yes*. Any job that exists today in 2018 will be automated, just like the almost all of the major occupations we had 200 years ago. The question is when— it could happen long after you and even your grandchildren are gone. But the bottom line is this: humans *cannot* be complacent about their skills and vocations. One must stay on top of a constant reinvention of oneself against the backdrop of relentless automation. That's a full-time job in itself. Welcome to the automation era.

Summary

Automation wins, humans lose. That is the popular notion that is being peddled by some corners of the media by so-called experts. On the face of it the statement is wrong: there is no precise zero-sum game with respect to automation supplanting human work. Productivity at the top line is a far richer reward for firms than labor cost reduction at the bottom line, and the pursuit of productivity places a greater emphasis on leveraging the existing human labor pool. This is where the vast majority of the best corporate thinking is positioned at the moment.

Moreover, automation can actually lead to much more fulfilling careers and interesting job roles. The best companies will provide strong incentives for automating one's own job.

We've addressed the "who" part of automation with this chapter, the next chapter will address the "where". Where will we build all of this automation? Where is the factory for new ideas, and how do we test them? The answer is … the *sandbox*.

Bibliography

1. https://seths.blog/2017/04/24-things-artificially-intelligent-computers-can-do-better-than-you-can/.
2. "U.S. Retailers on Pace for Best Holiday Season in Years", *The Wall Street Journal*, December 28, 2016.

12

Automation Sandboxes

In previous chapters you were introduced to a range of technologies for assembling automated versions of systems. I painted a picture of a culture of experimentation underscored by an architecture that selectively gathers a number of technologies all working in concert to achieve a degree of automation. The question arises: where do I do all of this work?

Enter the "sandbox" concept. In a real sandbox you have a walled-off area, scattered with toys allowing children to play freely. The sandbox is a fully outfitted maker-space for data scientists and analysts. A proper sandbox should have all of the toys (tools) needed to conduct any conceivable analysis. Here's how it works:

A secure "private cloud" environment is set up inside an organization completely behind the corporate firewall. This provides all the advantages of the seamless cloud architecture with no extraordinary compromises with security. The private cloud is completely separate but interconnected to the Production IT environment for data sharing purposes.

In actual fact you already have a sandbox in place, just not a very good one. Most firms assemble spreadsheet models based on data gleaned from corporate networks. Creative play happens when a user creates a spreadsheet to understand a corporate phenomenon better. Yet, spreadsheets are extremely limited in computational power, and, once done, are usually filed away on a random hard drive never to be seen again. That is an ineffective and inefficient way to gain insight and to contribute to the collective knowledge of the organization.

© The Author(s) 2019
G. E. Danner, *The Executive's How-To Guide to Automation*,
https://doi.org/10.1007/978-3-319-99789-6_12

In the cloud, on the other hand, users can create simple functions or sophisticated models using a language that is purpose-built for mathematical problem solving ... the technical computing language is in place to serve a wide range of computing needs that go well beyond spreadsheets.

A professionally done, purpose-built corporate sandbox is for those who are serious about automating their firms, serious enough to arm a group of practitioners with a specialized platform to do the work. There are as many opinions about the ideal sandbox for analytical and automation work as there are experts in the field. To narrow the question to a practical level I am going to make some assumptions about the kind of work that takes place in the sandbox:

1. The automation candidate starts with an idea about an algorithm that can automate an important function in the organization. That idea is expressed in the form of a diagram.
2. Our goal is to turn the idea on paper into a working piece of software.
3. We wish to easily interface that piece of software with other corporate systems outside of the sandbox to do real work.
4. We want to use a software language that is appropriate to technical computing of an arbitrarily complex function.

The factory (more like a garage) for producing and testing code is known as a Development Environment (DEV). Shepherding a coding project through the factory is called DevOps. Every DEV is a little bit different as styled for a particular language and purpose.

Over the years I have personally worked with every computer language under the sun, starting with BASIC on a NorthStar microcomputer in 1980 as a freshman in college. I was one of the lucky technologists by virtue of my birth year (1961) to witness the personal computer revolution from its infancy to its heyday and beyond. Languages came and went, with each new one promising to be THE language that would surpass them all and remain a permanent fixture of our landscape. None ever did, but instead what occurred was a fascinating evolution in the rich fabric we were able to use to glue an idea expressed in simple words and pictures to a living, breathing piece of code that encapsulates that idea. The hot language of the day added yet another handhold that carried forward to the next language of the day.

Mathematica from Wolfram Research, now known as The Wolfram Language, was well ahead of its time when it was released in 1988. Built from the ground up by human phenom Dr. Stephen Wolfram to be a technical computing language for scientists and engineers to further the cause of research and design in various fields, its unique notebook-style interface with a vast library of built-in functions set a wholly new course for analytical problem-solving. As the name suggests, a notebook contains notes, words, pictures that are journaled together in a handful of pages. The notebook interface mimics this style but interweaves code in just the right places to make the whole system self-explanatory. For the first time in history, a programming language was invented that allowed for tinkering and dabbling with ideas in an experimental fashion, born from the scientific culture from which Stephen Wolfram began his remarkable career. Thirty years later The Wolfram Language is still loved and admired for its ground-breaking design that has endured the test of time better than any computing language before or since.

Not long after the birth of *Mathematica*, a Dutch developer named Guido van Rossum released Python in 1991. From the very inception the philosophy of Python was as a "language for everyone" built with ease of use and intuitiveness as a governing set of principles. It was also open source, allowing both free distribution and community collaboration. Python promoted a concept known as functional programming, an approach that centers around compact expressions of functions as opposed to structured or imperative programming, which uses more primitive stepwise statements to build up a calculation.

Line for line, writing code in a functional programming style is far more compact than their imperative cousins, more readable, and easier to debug.

Python took its place as a solid general-purpose language but remained one of many in that category for several years.

Ten years later, in 2001, came a programming interface for the Python language called *iPython*, written by Colombian physicist Fernando Pérez. It contained a notebook interface much like The Wolfram Language where words, pictures, formulas, and actual working code could all peacefully co-exist on the same page, contained in one file. Pérez, an active *Mathematica* user at the time, originally built iPython to improve the quality and interactivity of scientific papers. It was also an open source distribution. This came along just at the dawn of Data Science as a profession, which propelled Python to the tremendous popularity that it enjoys today as a language of choice for analytical work.

Dr. Pérez was joined by a first-class team of developers and contributors to iPython, which continued to get better and better at a rapid rate. In 2014 the "project" was rebranded *Jupyter* as support was added for multiple underlying languages outside of Python. In many ways The Wolfram Language is technically superior to the combination of Jupyter and Python, but both share an important role in the automation sandbox together, as we will explain.

This profound evolution of Python, functional programming, and the notebook interface have colored my views on what I consider to be the best option for the core language of automation-directed algorithms. My philosophy is this: getting a solution to a problem is only half the work. The other equally important half of the work is *explaining* your answer to a skeptical or possibly even hostile audience. A language that allows you to do both in an elegant, efficient way is the right choice.

Python represents an entry point for a whole class of non-programmers. Let me explain: people who possess programming skills of any stripe are the ones with the most creative, most rewarding professions on Earth, and will remain so for the foreseeable future. I have made clear in this book that those who are not programmers have an equal seat at the table of automation and algorithm design as do programmers. However, let me be just as clear about this fact: if you are clever and successful and a budding systems thinker and you subsequently add programming to your repertoire of skills it will be as if you have now acquired a superpower that dramatically accentuates every other talent that you already had. I have seen doctors, CEOs, real estate agents, journalists, attorneys, musicians, engineers, and accountants in the late stages of their careers take up programming as a new skill—and their mindset is transformed. People like this suddenly have an invaluable technological outlet for their worldview—truly empowering. For the practitioners reading this, consider the sandbox not only your playground but also your classroom, should you choose to pursue this path.

Figure 12.1 represents our suggested architecture for a "Pythonic" sandbox.[1] Please feel free to use creative license on the pattern that is presented here—this is intended as a rough template for a sandbox in a world filled with technical options for each component. Note that this is not intended as a deeply technical discussion for a group of experienced developers, as I have deliberately glossed over many details in this architectural representation for practitioners who wish to understand the componentry of sandboxes at a higher level.

---

[1]We should note that the choice of other languages does not completely alter the architecture of the sandbox. You can flex this basic sandbox design in subtle ways to cover many other kinds of languages and platforms.

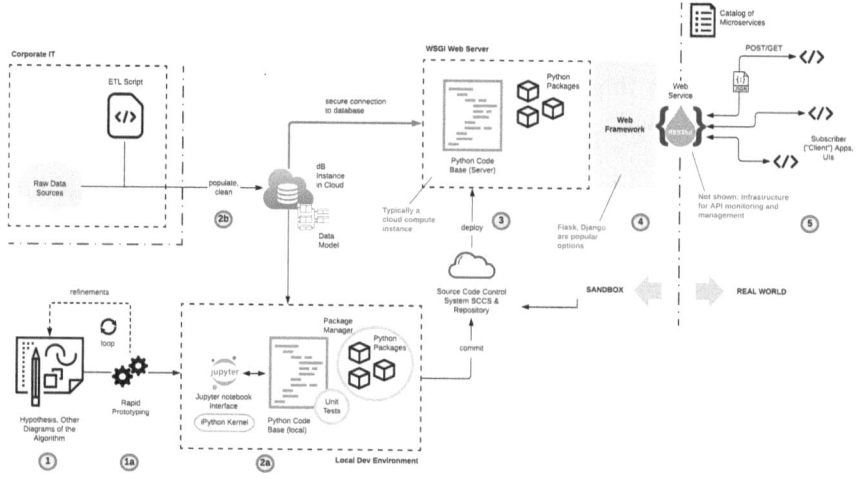

Fig. 12.1   Sandbox architecture

1. Diagrams

The sandbox is where serious developers play. Practitioners are likely those who have shepherded the automation candidate from a very rough idea floating in hallway conversations to a tangible concept for value-added application. You have followed the advice given in Chapter 5 on creating a series of diagrams that thoroughly tell the story of this function: the motivation, the logic, the data sources, and the goal. These diagrams will often form the handoff from the practitioner who is not a developer to a developer who is deeply skilled in the technical skills needed to turn the diagram into code (only in rare cases is this the same individual). The word "handoff" is not perfectly accurate here, as it is far more common for there to be a dense on-going dialog between the practitioner and the developer as the development stage is getting started.

It is important to keep in mind that these diagrams present the only window into the automation story for the developer. They must be comprehensive and succinct at the same time, not shrouded in industry or company language that is likely to obscure the technical goal of the function. Web applications for technical drawing like LucidChart are extremely handy for collaboration between practitioners and developers—the web format allows for comments from both sides to be applied to the diagram as development ensues.

1a.   Rapid Prototyping

It is often necessary to go on excursions from the mainline development to prototype certain parts of the system. In one recent project our company was asked to perform 50,000 identical calculations of a forecast for product demand. To reduce risk and uncertainty in our development we isolated one particular product and built the calculation for that single product alone. The result allowed all sides, including our client, to see and account for the challenge areas in performing the calculations. This "atomic unit"-level work is a very common frame for rapid prototyping.

Mathematica is a complete technical computing environment that is particularly adept in rapid prototyping in the manner I have described here. It is a desktop application that allows you not only to write compact code very quickly, not only visualize the results just as rapidly, but also create visual models that respond to user inputs like buttons and sliders with ease. These features conspire to make *Mathematica* the clear winner in the rapid prototyping space.

2a.   Development Environment

Our DEV example starts with Jupyter, a browser-based editing and execution app in the notebook format that we introduced earlier. Jupyter not only allows the developer to create Python code, but also document that code with a rich palette of textual features to the extent that Jupyter "notebooks" (files) as they are called are more similar to scientific papers than pure code listings from a text editor [1].

Like many open source languages, the vast Python community is constantly contributing code into the public domain in the form of packages that perform common functions like array handling and geographic visualization. Developers rarely build functions themselves if they are available in the form of a package free of license fees on the web. Package Managers are special applications that manage the constellation of packages that developers accumulate over the course of a project. Conda is an excellent package manager that is bundled with Anaconda, a popular distribution of the Python language core elements.

Developers will often use a test-driven development approach, which calls for the creation of Unit Tests in parallel with the main code base, promoting a healthy "build a little, test a little, build a little more, test a little more" cycle. Python packages are available for creating unit tests (which look just like code) within the project directory tree right alongside the code base.

Best practice suggests that developers save (commit) their project files as they are edited to a separate repository that maintains version history. GitHub is a popular service for this purpose.

2b.   The Database

Most algorithms use data to perform their calculations. A function that automates a corporate decision will draw on data elements from the company as appropriate—customers, transactions, assets, products, contracts— the molecules of data that comprise universe of information about the company. The data model, specific to the function at hand, represents the *structure* of the data in the database, designed by a developer proficient in data modeling techniques for efficient extraction.

The original source of the data lies somewhere in the Corporate IT environment. Stewardship of sensitive corporate data is of utmost concern to this group. They will insist on control of the process that pumps data from its original source into the database in the sandbox for your use. This process is known as Exchange Transact and Load (ETL) and exists in the form of a script that gathers the data together in just the right way to fit the data model at the target. Corporate IT will "own" the script that executes the data updates.

3. The Web Server

To expose the completed code to wider use, we must copy it to a web server (a server attached to the web) so that it can be "served up" to a browser-based audience and controlled via a communication protocol on the web. The careful and deliberate copy of files from the code repository to the right locations on the target web server is called *deployment*.

4. Web Framework

The code base makes use of special functions inside a package of packages known as a web framework. This framework allows the code to interact with other applications on the web through a protocol called a web service, the most common of which is a RESTful Web Service. The web service exists at a specific URL, and calls to the web service as well as responses are known as POSTs and GETs, respectfully. The data that is exchanged in these transactions is JSON files, a text-based, human-readable tree structure. Flask and Django are the two most popular web frameworks at the time of this writing.

5. Clients

Any application can send POSTs to the URL bound to the specific web service (we are setting aside the issue of security for the moment). A subsequent GET from the application retrieves the data resulting from the action of POST. In this manner these applications are referred to as "clients" of the web service, as the web service dutifully performs the action commanded of it from the client. That could include but is not limited to a rigorous User Interface (UI) or a visualization.

The good news and the reason that it is great to be alive and doing this work in 2018 is that all of these components of the sandbox, summed together, are ridiculously inexpensive. The vast majority of the elements are open source, and many firms already have some of these in place. Therefore the "ask" to the purse holders in an organization does not come along with a large price tag.

Moreover the uniform environment of the sandbox promotes code reuse. It is expected that users will take advantage of code fragments that have been created before, snapping them together in "Lego-like" fashion to enhance the speed of development over time and to reduce errors. Done well, the sandbox should produce more code per unit of effort over time. Culturally the objective is for the sandbox to become the "go to" place for exploring the "machine that builds the machine" as Elon Musk so eloquently described the science of operations excellence [2].

The sandbox as described here is the kind of architecture that blazes a path to production. At some stage a particular automaton rises in popularity and use from its experimental roots to become a critical app for the entire company to use. This is the point in the process where Corporate IT can be a critical linchpin of success, as they are well versed in the "productionization" of technology. In essence, IT will create a mirror image of the web server shown in the diagram and move to add tools to that production web server to balance the compute load, monitor subscriber demand among apps in real time, and perhaps strengthen the cyber defenses.

Why, you may ask, did I go to such elaborate lengths to describe the sandbox and to justify the elements that go into it? Because building sandboxes is politically hard work. As a practitioner and champion for automation you will be challenged left and right on both the idea of it and the choices you make. The IT department will claim that this is "non-standard" and "doesn't fit with our corporate guidelines" (I consider that a pro, not a con). Morons at all levels will suggest, "hey why can't you do this in Excel? I hear you can do a lot with macros". Technology purists will say that this language or that tool is actually better than your design for reasons rooted in banal minutiae.

My goal here is to arm you with clear and cogent talking points to factually persuade senior leaders that the greatest path to a better company is through automation, and automation is serious business that requires purpose-built tools. Anything less dangerously compromises that mission.

Summary

Successful automation requires focus. Part of that proposition includes a place well suited to the work at hand. An operating room is adorned with equipment arranged in a layout to facilitate good surgical procedure. There is no reason we should be any less dedicated in our pursuit of organizational value through automation.

The equivalent of the operating room in automation is the sandbox. There are infinite ways to structure a development shop for automation—here we have suggested one blueprint that has worked well in the past. Regardless of your ultimate design, the automation sandbox is the playing field on which the serious work of automation is done. Serious, yet playful, because another important aspect of automation is the experimentalist culture that allows it to thrive.

Once your sandbox becomes the living, breathing organism that it should, buzzing with activity, it will evolve into a vital corporate function, churning out new automatons regularly. As we will explain in the next chapter, it will become an *algorithm factory*.

Bibliography

1. "The Scientific Paper Is Obsolete", *The Atlantic*, April 5, 2018.
2. "Tesla's Quest to Build the Machine Behind the Machine", *Fortune*, June 6, 2016.

13

The Corporation as an Algorithm Factory

The purpose of a corporation in the automation era is to produce algorithms. Once produced, the algorithms act like machines that print money. All other functions in the corporation serve to support this basic purpose. The previous chapter described automation sandboxes in detail. Now that you know how they work, these sandboxes should be deployed as virtual production lines, spinning out useful algorithms repeatedly and frequently to be incorporated into the value chain of the organization.

Let me tell you the story of an algorithm "factory".

LeverJet is a company in the fractional jet business. It is a fictitious composite of several companies I have worked with in the past. LeverJet makes its money by offering flight hours on its fleet of private jets to corporate executives and wealthy individuals who purchase a membership in one of several plan offers. The company owns about half the jets, the other half are part of a pooling arrangement with other aircraft owners. Similarly, a lengthy pilot roster lists individuals based all over the globe, each with a unique set of certifications for various types of aircraft.

LeverJet was founded by a professor of mathematics at a prominent university, who built the first formal scheduling simulation model of the fractional jet business. The model became the linchpin of the business plan that was pitched to an initial round of investors. Today LeverJet is one of the leading firms, but in an increasingly crowded market.

Customers arrange their flights with a human coordinator. There is often a bit of give-and-take as the coordinator tries to stage flights using aircraft already in the area to avoid lengthy dead leg segments. Once the flight with

© The Author(s) 2019
G. E. Danner, *The Executive's How-To Guide to Automation*,
https://doi.org/10.1007/978-3-319-99789-6_13

dates is agreed, the scheduling system matches passengers with aircraft and pilot in the most efficient way. The system seeks to optimize the schedule by fulfilling the customer's needs at the lowest operating costs.

When LeverJet began operations 10 years ago, it was barely cash flow positive. A favorable jet fuel environment coupled with double digit increases in demand raised the financial performance of the company to a healthy level today. It turns out that corporations that had fleets of private jets on their balance sheets were eager to escape the optics of that "extreme luxury", but still needed the ability to quickly move executives across the country to airports that are typically underserved by commercial airlines, for example to visit plants in rural locations. LeverJet's success attracted many imitators, including flight brokers who owned no aircraft at all, relying exclusively on a 3rd party pooling arrangement with aircraft owners.

Two years in it became clear that LeverJet would not survive the inevitable imitation of its business model by simply incrementally improving the system over time. It needed step changes to its performance year on year, changes sourced from its world-class talent pool of former corporate fleet managers, pilots, and commercial airline experts. In a bold, strategic move it made the decision to create an algorithm factory.

In its first few operating cycles, LeverJet personnel came up with a long list of candidate improvements to the system. Some were implemented, some remained on the wish list. Making changes to the core scheduling system as the system was in full operation was painful and risky, and, unbeknownst to its customers, the LeverJet system went down on occasion in many cases due to the constant tinkering.

The Chief Operating Officer, we will call him Glen, came from a small but highly profitable regional airline. Before that stint, Glen spent over a decade in the pharmaceutical industry. Glen was very familiar with the experimentalist's culture ever present in pharma, and felt that this could work even in the "never fail" militaristic environment of the airline industry. He suggested that the company not only needed a way to quickly assimilate all of the best improvement candidates, but also to develop separately from the mainline of the up-and-running scheduling system. Glen became the inventor and executive sponsor for the automation sandbox that went live a year later.

In its first year of operation, the factory spun out three major algorithms. The first one looked at one of the biggest expense items on LeverJet's P&L: jet fuel. It just so happens that jet fuel prices vary widely across the country (and change every day), much like gas station prices vary between two stations a block away from each other. If the system could sense in real time that one of our aircraft happens to be in a low fuel cost city, it would forward this recommendation to the captain on duty for that flight.

The second application took a critical look at the pilot duty roster. Federal air authorities worldwide have very tight restrictions on how long crews can fly before a required rest period, on top of a longstanding list of rules around aircraft certifications. To make things more complicated, pilots are continuously looking for promotions by gaining certifications for larger aircraft. It is an insanely complex task to match a crew to a flight in a way that is compliant with all of the rules, so LeverJet employs 20 analysts that do nothing but certify flight schedules against duty roster and certification requirements.

The algorithm that LeverJet built took all of this into consideration. It let the flight coordinators know in advance about crew restrictions as they were working with customers to schedule flights. It also detected "inflection points" in the system where small changes would have a big impact, such as "if we just had one pilot based in Dover who could fly a Falcon 900 that would relieve a huge bottleneck" or "If we could bring one more G5 into the system in the Western region, it would make a huge difference to our growing customer base there".

The last algorithm in the first year wave pertained to product development. At its inception, LeverJet had only one product: a membership that entitled you to so many flight hours per year, with the hours priced on a sliding scale. LeverJet noticed that across all of its customers, profitability varied tremendously, with 10% of the customers at negative profit! The high costs came from flights that created dead legs coupled with costs for moving crews back to their home bases. Experts from the Marketing team put forward several ideas for offered plans that would alleviate some of these costs. One was the RegionCard, a membership that allowed flights within certain geographic zones only. Another was the City Pair card, a card that allowed flights only between two cities, perfect for wealthy families that had a home base and a vacation home or ranch that was frequently visited. In all, six new products were initiated that first year, all simulated and tested. These products had the dual effect of increasing customer satisfaction and demand while improving the operational performance of the company. The whole effort was so successful that an entire department called *Product Innovation and Development* was created to focus on the now active product pipeline.

A few years later the CEO (the former math professor) gave an interview to a leading business magazine. The interviewer remarked how interesting it was that the scheduling system he devised as an academic became the basis for a large, profitable enterprise and also formed the Genesis for an entire industry that had not existed before. He corrected his interviewer, saying this:

> While it's true that my scheduling model was the start of LeverJet, that's a small part of the story today. Perhaps what is most significant is the decision we made many years back to create a formal sandbox for innovation. Our people use it to devise new ways of working. Fractional jet service is constantly shifting, with new products and markets coming out every single day, so an innovation 'stack' that simultaneously improves our operating margins while keeping customers happy and loyal is really the core of the company now.

The LeverJet story is a classic example of an algorithm factory at work. In the prior era it was sufficient to be an asset caretaker—only a few firms had the capital to own expensive planes, so the competition was limited. Companies could simply coast along resting on a few incremental improvements here and there. Today, value is created primarily through clever algorithms that are built on top of a layer of assets, and increasingly these assets are not even owned by the algorithm provider. The game has shifted in this way for many asset-heavy industries from transportation to chemicals to construction to energy. The firms that will succeed will be those that make a science out of innovation, and back that up with a virtual laboratory that allows talented people to test a broad range of ideas for practical application. "Benevolent Hacking" is a phrase often used by our younger generation to describe this experimental culture that races to deliberately disrupt the current business model before outsiders do.

You may already have a pre-forged path to algorithm application without even realizing it. Starting in the early 2000s, Business Intelligence (BI) tools began to permeate the corporate market, products like Tableau, Spotfire, and Microstrategy. While BI is a fancy, highminded term, it really boils down to software that enables developers to create dashboards using a large gallery of graphs and plots, sometimes overlaid on maps. I have observed companies going a bit overboard with these applications, devising an overwhelming number of dashboards for everything from HR to finance to operations. I am a big fan of measurement and of the visualization of data, but it seemed as though we went too far "just because we could".

I was sitting with a client, a large retailer, as they were proudly showing me their dashboards. True enough, these things were strikingly beautiful, and the teams had done a great job integrating ERP data and other data sources to build meaningful portrayals of data, organized by logical themes. Finally as we landed on one particularly vivid dashboard entitled "Sales

Velocity Through Channels", I raised the question: "So what is this dashboard telling us?"

There was a notable pause.

They proceeded to explain each of the 5 or 6 plots on the page and what data they drew upon, and why that metric was important to sales.

Thanks for that explanation, but it doesn't answer my question. Say I'm on your sales team. What are these metrics asking me to do?

Another even longer pause.

Finally, one of the SMEs in the room spoke up. She explained that one of the channels showed a downward trend in velocity. That coupled with snapshots of a few other sales patterns on the page indicated that one of the dominant products in this category has likely reached the end of its brand life, which in turn implies that we should alert the buyer group that it is time for a seasonal refresh to the brand.

Of all of the regular viewers of this dashboard in the company, how many could have made that diagnosis?

"One" (the SME in the room).

The corporate world got so caught up in the zeal to apply Business Intelligence that they forgot the last mile to value: *interpreting the metrics on the dashboard*. I suppose many assumed that the diagnosis is obvious, given the right choice of metrics, portrayed in just the right layout. How many medical diagnoses are obvious when looking at 5 or 6 vital signs? If I sat you down in a commercial airliner and asked you to identify where the throttle was, where the altimeter was, fuel gauge, airspeed…could you do that? Yes, you probably could. But does that mean you know how to fly an airplane?

The SME in the room that day was brilliant. She was extremely articulate about sales and the underlying Calculus of how it all worked. She had numerous algorithms in her head mapping the metrics to a call to action. If we could capture those algorithms, we could run them alongside the pretty plots to create a superadditive combination—one that allowed every viewer to be just as informed as the best SME. This capability is commonly referred to as a Recommendation Engine.

For the practitioner, here is a tailor-made opportunity for highly effective HIL automation for those many companies that have enthusiastically embraced BI. All of the same automation processes we have discussed thus far apply just fine to the creation of Recommendation Engines to accompany each critical dashboard, with the text from the engine, clearly spelling out the call to action, appearing right alongside the plots:

Inventory of product X is running lighter than normal for this period of the year. Suggest setting the reorder point to 120% of current value.

The tank balance does not agree with the shipped volume. That means a leak in sector 6 or a possible failure of meter 799A.

The recent resignation of the sales manager in the Western territory is having a 4% downward effect on sales this quarter.

The demand for automation generated by BI is like having an algorithm factory with customers waiting in line outside ready to take the product as soon as it is made. For those firms that have embraced BI, the next logical step on the path is to create recommendation engines that run alongside the dashboards providing the same advice that a human expert would if he or she were standing next to the viewer.

In a very similar vein, companies that begin the move onto the blockchain are paving a path to automation whether they realize it or not. The blockchain is not just a universal distributed ledger for recording all business transactions and assets. It also provides the mechanism for smart contracts that we discussed in Chapter 6. Smart contracts are devised to execute a pre-defined agreement among various parties. Any modern company these days has hundreds, perhaps thousands of contractual agreements with customers, suppliers, and partners. Some are formal, others implied. If you were to construct an inventory of all contracts for a given company (as we once did for a division of an oil field service company), you would find that the vast majority are manually administered, casually enforced, poorly monitored, and frequently violated by all parties. A smart contract would counteract all of these shortcomings while making the administration and self-reporting of its status completely automated. Being part of a blockchain means having ready-made infrastructure for building smart contracts. The same process for general automation work that we have described thus far applies just as well to the building of smart contracts, perhaps the only difference being the need for an attorney as one of the SMEs on the team.

There is a deliberate meaning to my use of the word "factory" in the title of this chapter. A factory produces output continuously. There's never a question about the factory running—it is what allows the business to earn value. No factory, no revenue.

Yet, when we look at automation and the building of algorithms to support it, often a "project" mentality takes over. We engage in a series of one-off projects to automate this and that, and many of these projects don't even get off the ground as they fail to make it through the approval process. It is a completely different way of thinking from the factory imperative.

The best firms will take automation seriously. That implies an expectation that the automation program is "always on", churning out products (automatons) on a regular basis, emanating from an efficient sandbox. It also implies that defects and design flaws in these products are caught quickly and resolved before they reach the customer. The work of dedicated automation never ends.

Summary

Automation is not the latest management fad. It is the fundamental transformation of business from its current, largely unautomated state to one that operates with far less human direction per unit of output. It is a relentless, unyielding march—not something to be taken lightly. As algorithms are the molecules of automation, organizations must turn themselves into factory-like mechanisms for producing vast collections of these algorithms in a systematic way, using the sandbox as we described in Chapter 12. Once built, these algorithms comprise the lion's share of the organization's intrinsic value.

Faster/better/cheaper is the plain vanilla payoff for firms using automation. Beyond this, organizations have myriad ways to make money directly from their collection of algorithms. In the next chapter we will take a tour through these value development steps so that you can realize the best return on your investment in automation.

14

Monetizing a Tapestry of Algorithms

Algorithms that automate such as those devised by the LeverJet team are only of passing interest if they are not put to work, harnessed for value. For LeverJet, the value was obvious, because each algorithm had a direct bearing on the company's operations and customers. The value might not be quite so obvious for other organizations with different characteristics. Can not-for-profit enterprises monetize algorithms? Can investors participate in the algorithm development sub-economy? What about start-ups?

The answer is yes, yes, and yes, and this chapter will explore the avenues to value for all firms, including those that do not have quite the same apparent value proposition as we saw with LeverJet.

If you are like most firms, long before this book you had no formalized process for creating algorithms, so they rooted where they stood—in people's heads, in spreadsheets, in emails and reports. Perhaps they remained "ways of thinking" that floated through the hallways never actually meeting pencil to paper. Even firms that went through the "Knowledge Management" fad of the late 1990s really never achieved their goals of cataloging the enterprise intelligence in an effective way.

Here is my suggested plan for a practitioner:

1. **Navigate through the enterprise to discover algorithms wherever they lie**

This can start with the simple question: how do we do X? You would be surprised where that initial answer leads, often to algorithms that sit at the

© The Author(s) 2019
G. E. Danner, *The Executive's How-To Guide to Automation*,
https://doi.org/10.1007/978-3-319-99789-6_14

very core of what makes the firm successful, yet, has no formal recognition whatsoever.

Algorithms are not always tied up in a nice, neat package with a bow around it. In the real world they are messy and fragmented across people and machines with the integration points even less clear. To counter this you have at your disposal the drawing tools as revealed in Chapter 5, an excellent starting point for putting a logical lasso around the boundaries of the algorithm. Just as helpful is the Digital Twin concept that was introduced in previous chapters. When you begin to think of a Digital Twin of a human expert, a group of humans, a machine or a collection of machines, the thought exercise that this one simple phrase brings to mind is often helpful as an organizing framework for capturing algorithms.

2. Create a trackable inventory of algorithms

A modern organization wouldn't even think twice about keeping a list of all of the assets it owns, along with a constellation of meta-data about each asset. The same should hold true for valuable algorithms. Such a list should be published throughout the company, updated on a regular basis, reviewed in board and strategy meetings, and be constantly expanding. If the inventory is *not* growing, what does that say about the firm?

3. Prove (through models) that the algorithms generate business value

Remember the ROI models for the CFO that we mentioned in Chapter 10? These can and should be applied to individual algorithms to show their value. In fact, for an algorithm to be a genuine Digital Twin, it must have in its make-up a model that self-reports its value. If the value is increasing or decreasing over time, this is a helpful metric to know, and might imply changes to the algorithm or altogether new algorithms needed in the mix.

The accounting profession has struggled in the last generation to value intangible assets like algorithms, even if those algorithms drive the value of the firm. How much is the value of the formula for Coca Cola? The patent for speech recognition? The know-how to build a bridge? The only way accountants offer up value measurement for these things is to use the residual method. The sum total of all intangible assets is categorized as "goodwill" on the balance sheet after all the hard assets are subtracted from the market value of the firm. This strikes me as woefully inadequate for this day and age, much less wholly unsuitable for a highly automated future.

Take two firms that make exactly the same product. One is highly automated, the other is not. With the financial measurement tools we have today, odds are that both firms will come out with roughly the same value, yet we know logically that the automated firm is more valuable, and the market is very likely to pay a premium for it over its unautomated peer. For firms looking to acquire other firms, the stock of inherent algorithms driving automation in a target to be acquired conveys a material advantage to the acquirer, not the least of which is the potential for scaling that company's automation acumen to the whole enterprise. When Google acquired Motorola in 2012 for $12.5B, it did so largely for Motorola's vast inventory of communications-related patents. According to a 2015 study, intangible assets now account for 87% of the value of firms in the S&P 500 [1].

Company leaders use managerial accounting principles to inform them of the value of their firms at any given time, and also to value individual resource allocation decisions that are made in any given fiscal year. How can these leaders make informed decisions if the chief component of value does not even appear in the equation?

Help is coming. A small cadre of leading academics in business accounting are keenly aware of this problem and are working toward its resolution through research. As automation practitioners it would suit you well to stay close to developments from the likes of Dr. Baruch Lev of NYU, who has written extensively on this topic [2].

4. Protect the company's property rights to the algorithm

In my public remarks on algorithms I get this question all the time: if I come up with a clever and unique algorithmic way to solve a hard problem or to automate a challenging process, can I patent that algorithm? I wish there was a clear answer "yes" or "no", but rather the answer is a little murky. My objective here is not to dispense legal advice and my strong recommendation is that you consult a patent attorney whose job is to stay on top of this very hot and dynamic issue.

In a landmark case in 2010 called *Bilski v. Kappos*, the Supreme Court sought to clarify the test for whether an invention such as a business process is eligible for patentability [3]. Many consider this to be the defining case for both software and algorithms as eligible or not. The key word here is *abstract*. You cannot patent anything that is abstract (like an algorithm), rather, you have to close the loop and pair that abstract thing with the actual application of the thing. As an example, an algorithm as a series of simple steps is not patentable, but when embedded into a machine that transforms some input to a different output, that combination is deemed patentable by the *Bilski* decision. It is a fine distinction that a good lawyer can advise.

Short of a patent, however, there are many alternative ways to protect an algorithm:

(a) Store the algorithm securely with access both recorded and limited to certain individuals.
(b) Register it with a 3rd party, like a documented record with your law firm.
(c) Compile the code in such a way that the human-readable form cannot be reverse engineered easily.

5. Formally register the algorithm inventory with the company assets

We are very careful with our tangible equipment. There is a whole category of software called Asset Management Systems where everything from office copiers to machine tools are registered in precise detail. We should treat our algorithms with the same degree of care. A simple database will do: list each one by name, the authors, the associated diagram(s) and source code if available. Make this an active repository, refreshed as frequently as its hard asset cousin. Recall the conversation on Digital twins in Chapter 2? These form a good organizing principle around the concept of a singular asset that is more easily valued than an amorphous collection of disparate technologies and methodologies.

The overall message here is that algorithms are important and will rise in importance throughout the automation era. Because of this, they should be respected enough to treat them as the valuable asset that they are.

There are stories of famous algorithms throughout history that have fundamentally transformed companies in profound ways. The lessons of these developments are instructive for us today. Let's examine a few of these.

Yield Management at American Airlines

In 1972 Ken Littlewood, a researcher at British Overseas Airways Company (now British Airways) devised a revenue model for choosing between one of two possible fare classes for the same seat based on the remaining unsold inventory of all seats at a moment in time [4]. He published his findings in a paper entitled *Forecasting and Control of Passenger Bookings*, which gained popularity among a small but growing group of specialists in this area. His model became known as Littlewood's Rule. A few years later in 1978 the

US congress passed the US Domestic Airline Deregulation Act which led to an explosion of small private carriers seeking to compete with legacy carriers. American Airlines responded to the threat by commissioning its own group of researchers to develop competitive pricing formulas that could be integrated with their existing reservation systems. It is not insignificant that this paralleled a dramatic rise in the adoption of mainframe computing power shepherded by IBM and "BUNCH", the nickname assigned to mainframe hardware companies Burroughs, UNIVAC, NCR, Control Data, and Honeywell. American devised a group of algorithms called Expected Marginal Seat Revenue (EMSR) that built on Littlewood's rule. Today, EMSR is still the basis for most Revenue Management (RM) systems that handle everything from hotels to rental cars to apartment buildings.

Robert Crandall, CEO of American Airlines, actually coined the term *yield management*. He credited yield management for generating $1.4 billion in incremental revenue for American over a three-year period when it was first introduced.

A great narrative about automation in its own right, this story also highlights another important point: defensive automation. Within a few years after American introduced yield management, all of the major airlines were using it in some form in order to survive. If you did not use yield management, you were on the outside looking in and your odds of survival were slim to none (a host of smaller airlines went out of business or were acquired in the 10 years after deregulation). Many airlines who were not American adopted this special form of automation as a defensive strategy, and a few tried to innovate on American's models. This "arms race" will not be uncommon in the coming automation era, which is why it is so crucial for all firms to develop a sandbox capability and human sleeper cells that can quickly spring into action as the market evolves.

The ORION Routing Algorithm for UPS

James E. Casey founded the company that eventually became United Parcel Service in 1907. In leading the company for the next 55 years he implored the firm to "to constantly seek better, safer work methods". It is not surprising that UPS went on to develop the On-Road Integrated Optimization and Navigation (ORION) system for over a decade, spending in excess of $1B in the process. ORION went live in 2013 at 10,000 company sites and completed the rollout in 2016. It has been called world's largest operations research application.

ORION uses a combination of road data and live GPS feeds to optimize delivery routes for its drivers, based on fuel consumption, drive time, and distance.

It has been estimated that the presence of ORION versus a more simplistic route plan saves UPS about 100 million miles of driving per year. A quick calculation reveals that the direct savings swamps the "paltry" $1B investment in the algorithm, not to mention the safety effect of fewer miles on the road.

LinkedIn—The $26B Database

LinkedIn is a very popular service, primarily for professionals to share their contact information with a trusted group of colleagues online. The system somehow knows magically that a connection of a connection of a connection of mine and I used to work at the same company and might know each other (and therefore suggests a direct relationship). This kind of connectivity intelligence is made possible through graph technology of the same type we discussed back in Chapter 6. In essence, LinkedIn is a gigantic graph database that the company refers to as a "knowledge graph". All of the services that LinkedIn provides (job search, targeted ads, lead generation tools, content posts), and all of the revenues that it generates are built from analytics that are performed on top of this graph data structure. Engineers at LinkedIn say that its business model would not be possible without the graph nature of the underlying technology.

In 2016 Microsoft purchased LinkedIn for $26B, making it perhaps the most expensive database in the world.

The Amazon Go Retail Model

Amazon is a remarkable firm, not just for its financial success, but for its breadth of businesses. Like a supremely talented actor, Amazon has proven it can excel in industries as broad as retail, cloud computing services, and streaming media. Not everything Amazon touches turns to gold, but their track record is darned impressive.

Into this innovation hotbox comes the disruption of the grocery industry. From one single pilot near its headquarters in Seattle, the company created the Amazon Go store. Swipe your smartphone at the entrance, grab your products from the shelf, and walk out. Check out and billing are completely automated. Amazon correctly determined that the biggest non-value added queue time for customers (and therefore the biggest impediment to revenue velocity) was the wait for a cashier or a self-checkout lane. So, they simply eliminated it through automation. It remains to be seen if their subsequent acquisition of Whole Foods will see the Amazon Go model imparted upon it, but one could certainly draw that logical conclusion.

One of the algorithms that makes Amazon Go possible is known as *sensor fusion*, a phrase we introduced in Chapter 1. Sensor fusion as the name suggests seeks to draw a conclusion of fact by deploying multiple sensors at one time. A consumer removes a bottle of ketchup from a shelf. Some cameras conclude that the bottle is red and has a vertical elongated shape. Weight sensors on the shelf determine that the bottle weighs 32 oz. Looking at the history of purchases from this customer we see that every month they buy the same brand X bottle of ketchup. All of these disparate pieces of evidence add up to a 99% chance that the product is SKU 74639297, which today costs $1.99. Why not just install an RFID tag to each and every item in the store to get an exact match? The cost of RFID on every item would be prohibitive, but sensor fusion performs almost the same function at a fraction of the cost per unit. The 1% error rate is within an acceptable tolerance and is declining with each new generation of sensor fusion techniques.

Amazon Go is a great story about automation. But as we speak here in 2018, the story continues to grow. We know of several industrial firms who have been inspired by Amazon Go to consider the adoption of this model to commercial transactions. Say you have an automotive plant sitting in the midwestern part of the US. A countless cluster of smaller firms in a 100-mile radius of the plant provide parts and kits for the plant, shipped in by truck. What if both ends of the shipment—pickup and dropoff, were keyed from the driver's smartphone, opening the security gate, shining a colored light on the product to be picked up, verifying the load, all to be done in reverse order at the plant on arrival. Such a system would cut down on the endless delays of security gates, paper signing, driver waits, locating shipments—all of the "friction" that is built up in traditional unautomated supply chains. Amazon Go is a pioneering concept that will find application in all sorts of industrial businesses.

Disney's Theme Parks

Innovation is deeply engrained into Disney like no other company on Earth. It stands to reason that Disney is front and center on the use of algorithms to extract greater value from every business that they own, from Pixar to ESPN. An entire book could be written about all of the ways that Disney uses clever ways to improve their products, so for our purposes here we will focus on just one narrow aspect: managing their theme parks Disneyland and Disney World.

Disney has a long history of intelligent design, coming largely from its R&D arm, Imagineering. To enhance the guest experience at its theme parks, Disney works prodigiously on the least favorite part of the visit: waiting in line for a ride. Disney has whole teams of experts in queueing theory, the math behind waiting in line. Models assist in the design of rides and the management of queues in a way that minimizes the total wait time, even when the park is near maximum capacity. Disney is constantly thinking of ways to carefully balance guest traffic among all of the attractions in its parks to cut down on the biggest complaint—you guessed it—long wait times for the most popular attractions.

It was in this setting that Disney rolled out MyMagic+ in 2014, a $1B investment in an automated system for allowing guests to make pre-attraction reservations [5]. It also made use of guest wristbands with RFID so that their location was constantly monitored. Because the average wait times are posted as the guest makes their reservations, guests will naturally choose time slots for those attractions in the lower wait time range, if they can. The system also allows reservations and ordering through the app so that the dining experience time is minimized as well, and Disney gets an advance warning of what meals are needed where. The end result is the removal of all of the "friction" in the park experience—great for guests, and valuable for Disney as they increase the visitor count for the same physical footprint.

These stories highlight the value of clever algorithms, well before the coming wave of automation. It was the company's choice to make a science out of something that was not necessarily considered a science at the time but turned out to be a highly valuable move. Even in the absence of automation, the science of being better is always a worthwhile pursuit.

Automation and its focus on clever algorithms are likely to help accelerate the trend in unbundling previously monolithic industries. For example, if you wanted to be in the automotive business you could not simply design cars on paper, rather you had to build and own a very expensive production system. Algorithms that form the basis of automated features inside the car can now be developed separately if the physical car actually becomes just a platform to house features. Let's say I come up with a very clever algorithm for more comfortably controlling the air conditioner in cars near coastal regions with humid climates. I could sell my algorithm to the platform company or directly to consumers assuming the platform was "open architecture" in nature (watch for this to come to the automotive market soon). Perhaps at some stage Ford, Toyota, Tesla, or BMW decide that owning factories is a misuse of capital and become designers and integrators instead, leaving the production task to a host of 3rd party (automated) factories that can make anything from a pickup truck to a sports car.

3D printing will serve as a tailwind behind this trend. If I am a designer of beautiful, ornate chairs I could distil the design down into a set of instructions for a 3D printer to produce. The value of my offer is in this design, presented as images in a web-based catalog. I sell you an instance of my design that you as a consumer take to your local 3D print shop (there will be as many of these are there are copy shops today) to print. No big factory was involved, no truck transport, just a designer-to-consumer interaction with the 3D print shop forming the physical last mile. Think of all of the industries that could be disrupted in this way: fashion, home construction, toys, tools, and jewelry to name just a few.

Unbundling production from design is a trend I have observed for the past decade in companies like Apple, and I expect it to continue to permeate other industries, albeit more slowly. This in turn presents opportunities for all kinds of design-oriented firms to monetize algorithms without the roadblocks of massive capital outlays for production facilities getting in the way.

Summary

Algorithms are the substance of automation, but it is important to recognize that algorithms in and of themselves are incredibly valuable. Managed well they become an asset right alongside other hard assets like buildings and machinery that convey value as well. The most successful firms will formalize their algorithm asset holdings, as other famous examples have shown us. This will in turn dramatically raise the market value of these companies, even in the absence of a recognized way to measure the value of such an intangible asset (coming soon).

Up to now I have painted quite a rosy picture of this automated future before us, which genuinely reflects my belief that automation on net will be good for our organizations, our workers, and our society. But that does not mean that things won't go wrong along the way. They will. The next chapter takes an unflinching look at the way automation could go wrong and offers thoughts on counteracting the negative effects.

Bibliography

1. http://www.oceantomo.com/blog/2015/03-05-ocean-tomo-2015-intangible-asset-market-value/.
2. Baruch Lev, "Intangible Assets: Concepts and Measurements", *Encyclopedia of Social Measurement*, Vol. 2, 2005.
3. https://en.wikipedia.org/wiki/Bilski_v._Kappos.
4. https://en.wikipedia.org/wiki/Littlewood%27s_rule.
5. "A Billion-Dollar Bracelet Is the Key to a Disney Park", *New York Times*, April 1, 2014.

15

What Could Possibly Go Wrong?

What could possibly go wrong? Plenty, actually. Automation has its vocal critics these days from labor unions to economists to politicians to ethicists to safety advocates. And not all of them are wrong. Automation does have the potential for physical and economic harm if not implemented well.

I'm not an economist or sociologist. I have no credentials to legitimately weigh in on the *ethics* of automation. I don't have the skills to grasp vast societal and economic changes. Rather I'm a scientist of business, fascinated by the possibilities presented by automation to create wholly new companies from the assets of the here and now. If it were up to me, we would automate anything and everything because…we can, and it's fun doing so for those who are skilled in the domain. But we know instinctively that there are limits to automation. At the very least we can examine objectively the ways automation can go wrong, then let well-trained professionals respond to each case.

Cyber Attacks on Automated Systems

Attacks on technological systems by bad actors are a permanent feature of life today. Any website these days is attacked thousands of times a day, only a tiny fraction of these attacks are successful, yet the rare successful attacks are the ones we read about in the news media. It remains an unsettling environment to think that the least little vulnerability in our systems could result in a breech that allows an army of barbarians through the fortress gates. But this is the age that we live in.

© The Author(s) 2019
G. E. Danner, *The Executive's How-To Guide to Automation*,
https://doi.org/10.1007/978-3-319-99789-6_15

Automation is a special class of technology. By its very definition it is harnessed for some specific task, one important enough for the organization to spend time and resources to create it. If such a system were commandeered by an actor with ill intent the result could be disastrous.

The simple solution to cyber attack is to wrap any system in any one of the excellent the cyber security technology solutions that we have at our disposal today. Yet you are likely already doing that if your IT function is doing its job properly. But beyond the simple fix, are there design aspects of automation that we as practitioners should consider to harden them against attack?

Ensemble modeling is a design approach that seeks to create many equivalent models that "vote", committee style, on final decisions. In automation this has the effect of creating not one but several Digital Twins all in one package, each with an independent decision-voting authority but using the same algorithm, parameters, and input data. If you had a Digital Twin flying a commercial airplane, for example, no one of them could decide to go off course and dramatically decrease altitude. Not only would this Digital Twin be outvoted, but it would likely be rendered suspicious and taken off line by a vote of its peers.

In short the best answer that I can provide to the challenge of cyber attacks is that for every offensive challenge to our systems we *devise yet more technology and automation to the defensive side.*

Mass Unemployment

In my extensive research on this issue I have found no evidence in history of events where permanent, mass unemployment was created due to technological advances. Temporary job displacement, of course—telephone switchboard operators, blacksmiths, travel agents—all of these job categories were disrupted and humans left these positions in droves, and that will continue to happen in the age of automation.

But permanent, "mass" unemployment of the kind that noted authors are peddling seems very unlikely. Take the fast food industry, which presently employs around 3.8 million workers in the United States [1]. Let us say that a key firm in the industry embraces automation in a manner that allows a store to operate completely unmanned. How long would it take to roll out this automation to all of the stores in its system? McDonalds, for example, has about 37,000 stores worldwide. This is a 5-year undertaking in the best of scenarios, well enough time for workers to understand very clearly that this job category has reached its useful life.

Wealth Disparity

What happens to the people who fall below the line? The line in this case is the threshold below which machines can easily do their work. What happens to them? I'm no macro-economist, but in my own work and observation I've seen these kinds of folks get "promoted" to machine coordinator.

Let us return to our example of a worker at a fast food restaurant preparing individual meals on demand to customers that enter the store. That same worker in an automated setting might monitor the processes of an entire restaurant, perhaps even from a remote location. When something breaks, or a customer complains, the human orchestrator swings into action on an exception basis. The worker becomes a creative problem-solver, rather than a human gear in a large faceless machine. Does this require altogether different skills? Absolutely it does, and I for one would like to see our education systems take heed and slant their curriculum toward universal problem-solving/troubleshooting skills versus "book knowledge" alone. Mandatory programming courses in an elementary language would be helpful as well.

Re-skilling is the only sustainable antidote to wealth disparity that moves society forward. There is little doubt that some job categories that exist today will go extinct soon. I have no answer otherwise to those workers who fail to adapt.

Automated Harm or Terror

Can bad actors use automation against us? They already do. In 2013 the tragic bombings at the Boston Marathon were made possible using two pressure cooker bombs that were remotely detonated, killing 3 people and injuring 264. One could easily argue that the terror system in place that day was automated to the extent that the terrorists could engage the weapon from a distance.

The specter of a motivated terrorist commandeering a large automated system to repurpose it in real time to do harm is the nightmare scenario that we all fear. Large industrial plants, tractor-trailers, power generation, aircraft—all of these could in theory create great harm if under the control of bad actors.

I say "in theory" because the practical challenge of assuming control of a large system is substantial. It is made even more substantial by hardening the physical and logical perimeter around the system controls and at the same time distributing multiple copies of the controls so that a rogue-ly operating system is detected by its peers and deactivated.

In principle the protection of our automation systems against bad humans lies in ever more automation applied defensively.

I am more than a little embarrassed to write about the *Terminator*-like scenario, where automation somehow becomes self-aware and turns on its human counterparts...but I am forced to. Elon Musk, Bill Gates, Ray Kurzweil, Stephen Hawking and others have given very public statements about the dangers of superintelligence. To be fair they were talking about AI, but they could just as well have been discussing automation.

Sure, machines can be harnessed by humans to do evil. A well-placed virus can render a power plant inoperative or steal your identity. But these are bad human actors using technology as leverage. What these luminaries talk about is the *inherent* danger of automation, that in and of itself it forms bad intentions for humans. I have studied the writings of these naysayers for a year now, and I can't find a cogent technological argument that shows exactly how evil intent can arise from something that was never originally programmed as such. And even if the machine could self-generate knowledge and reason, isn't that just as likely to be benevolent than malevolent?

Unintended Consequences

History is littered with examples of unintended consequences of technology and machine-driven decision-making. Recall the out-of-control pricing algorithm used by Amazon UK in a previous chapter? The answer in that case, and in many cases of unintended consequences is a broader testing regime that considers the automaton in place, surrounded by a realistic replica of its environment. I will add that Systems Thinking is practically tailor-made for the purpose of surfacing unintended consequences and in fact has been used for that very reason in the past.

Over-Reaction

There are corners of our society that fear increased automation. In actual fact this has been true since before the Industrial Revolution. The popular term "luddite" came from a group of English textile workers led by a symbolic character named Ned Ludd who set about destroying new machinery used in the production process as it threatened to devalue the skills they had developed over many years. The movement ended after the British Army suppressed the violence, and Parliament in 1812 decreed that the destruction of a machine was a crime punishable by death.

These days we do not have organized groups going around bashing machines. What we do have are people of the same cloth seeking to tax automation under the label of the so-called "Robot Tax" [2]. Of all the bad ideas on the bad idea list, this one has to be the worst.

On paper, the idea is to create a general tax on automation that in turn funds a basic, guaranteed income for workers that are of a class that is likely to be automated away. You might see how that sterile logic might sound attractive to younger voters in lower skilled jobs, swayed by a young, articulate politician who obsesses about wealth inequality.

But there are several dangerous consequences to this proposal. First, it slows the adoption of automation at a time when it is so sorely needed to increase productivity, and in turn GDP for any given country. Second, the incentive to work hard as a basis for economic benefit allocation is broken with a Universal Basic Income, which will create unfathomable social consequences. Moreover, I do not buy the argument that wealth inequality is inevitable here—asset owners still must rely on automation producers for their automation, and those producers themselves are employers of many layers of trades. History has taught us that well-meaning interventions of government in the regulation of technological progress has been a rather blunt instrument. I advise against any taxes (both assessments and abatements) related to automation.

Institutional Instability

As consumers, we love disruption. When innovations like FedEx, Wal-Mart, Amazon, Southwest Airlines, Uber, and Airbnb came along we snapped them up with abandon. Better choices at lower prices—what could be better to a consumer? Yet each of these disruptions came along relatively slowly, at least slow enough for the legacy providers to react and adjust. Some did, some fell by the wayside. The best market proposition survived, which is the way it is supposed to work in a free market.

If the automation trend dramatically accelerates this disruption, we could find ourselves with the pendulum swung so far the other way that no company is able to maintain a dominant position and upheaval is around every corner. That is not a good situation for employees, consumers, and investors alike.

I have my doubts about the likelihood of this extreme case scenario, as I do believe that automation represents another business competency like any other—manufacturing prowess or great design talent as examples. Those firms that are really wise will in fact develop an advantage that could last for decades, providing a stable platform for great products, a rewarding place for employees to work, and a source of long-term returns for investors.

Liability

If a machine not under the direct supervision of a human hurts or kills another human, who is at fault? Nowhere is this question more front-and-center than in the current debate over liability for accidents caused by self-driving vehicles [3].

Software will make the decisions in cases where, say, a pedestrian jumps in front of a car. The car may actually choose to put the passengers at greater risk to save the pedestrian, or vice versa. Who is to blame for the subsequent injuries?

Most legal experts are acknowledging that the car manufacturer will assume product liability in the cases where system cause harm, shifting the liability away from the consumer. Manufacturers surprisingly have expected this shift and have not been seen to pursue waivers and other special treatment by standing tort law. Still, sorting out liability in the age of automation will be complex task, but I do not see that it will impede the development of more automation. In fact, on net there are two benefits to automated systems in liability cases. One, most automated systems capture reams of real-time data about their operation, so reconstructing the precise logic of what happened leading up to the accident is much easier. Second, since the vast majority of accidents of any kind are human error, it is believed that many accidents will be eliminated altogether.

Summary

Automation is of equal bearing to every other punctuation mark in our economic history: things mostly go right, but others go wrong, creating winners and losers at each stage. However, like watching a car accident happen in slow motion, we have the luxury at present to contemplate the challenges and develop policies and other countermeasures that minimize the negative impacts while accentuating the positive. Ethicists, economists, and sociologists are specially trained to examine human/technological dislocations and are invited to do so now, *before* we begin to see these effects on a massive scale.

No matter what, one fact is inescapable: our education systems are woefully unprepared for the talent requirements of the automation era. Programming is not a compulsory course in most Western societies; problem-solving as a skill is not recognized as a legitimate part of the curriculum. It is my sincere hope that leading educators will read this book and consider the ever-widening gaps between today's teaching content and the needs of the automation era.

It is my belief that on balance, in spite of these challenges, the net effect of the automation era will be a positive one, just as the Italian Renaissance and the Industrial Revolution were before it. That belief is based on my microcosm observations of automatons placed inside individual firms, making the firm better of course, but also enriching the lives and careers of the humans in that sphere. And generally speaking, the answer to most of the objections to automation is … more automation. Apart from my own view however, one fact is certain: we could not stop the relentless march of automation even if we wanted to. Automation is inevitable.

This has been a breathtaking journey across an exciting story that is unfolding before our eyes. A very different future is at our doorstep, and leaders in this era must personally prepare for its consequences. Over to you now, dear reader. It's your turn.

Bibliography

1. https://www.statista.com/statistics/196630/number-of-employees-in-us-fast-food-restaurants-since-2002/.
2. "Robot Taxes Are a Good Idea as Long as the End Goal Is Basic Income", *Fast Company*, October 10, 2017.
3. "When Driverless Cars Crash, Who Gets the Blame and Pays the Damages?", *Washington Post*, February 25, 2017.

16

Your Turn

People inside of notable organizations, like many things in nature and life, follow a Power Law distribution. In other words, a tiny number of people push the organization to become the thing that it is, good or bad. We are not talking about just the leaders at the top of the pyramid. Anyone can be a driving force.

I am going to assume that your motivation to read this book was to arm yourself with the skills to become such a person, or to find a direction for your energy if you already are. Congratulations, you've found the right place. Those who master the skills array of automation will be among the rare talents we look to lead organizations through this epic transformation. Wherever you are now, you will find yourself rising in responsibility.

In all of your efforts to make companies better/faster/cheaper (and to grow), always do so with a mind toward *ingenuity*. That's a word with a special meaning. It implies intelligence and sophistication, but at the simplest possible level. Some of the greatest thinkers and inventors of our time—Ben Franklin, Steve Jobs, Richard Feynman—all had a knack for breaking down complex bodies of knowledge into a set of simple but powerful concepts of design. A smart person builds a complex machine. A *genius* builds a machine that performs a complex function using piece parts that in and of themselves are individually simple.

In fact, I will venture to say that most of the things we observe in business and in nature have simple underlying roots. If you were in a helicopter hovering above a large city observing traffic patterns in an out, the complexity of it all would belie the fact that as drivers we operate by a relatively simple

© The Author(s) 2019
G. E. Danner, *The Executive's How-To Guide to Automation*,
https://doi.org/10.1007/978-3-319-99789-6_16

set of rules. Yet, when we are combined with many thousands of other drivers the patterns at the helicopter level seems incredibly ornate.

In my many years of solving business problem after business problem, I am always struck by this paradox. It gives me great hope that even the most complex problems that we encounter are tractable if we apply the right process, a process I have outlined in considerable detail in the previous chapters. It is also the reason I have placed such a persistent emphasis on the creation of drawings. Pictures are the language of automation, and the very act of creating a picture with notations itself yields insights by slowing down our thinking into methodical contemplation and abstraction. Leonardo Da Vinci and Richard Feynman were undisputed polymaths of their day for this very reason. They made drawing the observed world a habit. We would do well to adopt their habit in our own work.

It is perhaps this attribute that will separate the great practitioners from the good ones, and hence the great firms from the good ones: the ability to quickly characterize a business problem into a meaningful diagram that many others can follow. The technology comes afterward.

It is easy for us as practitioners, particularly those with a programming background to become enchanted by the technological aspects of automation. I understand the allure—the toolset of automation represents a breathtaking array of the very latest developments, and even veteran technologists are amazed by the tools and their power. And while I want to caution you about the seductiveness of the technology and the careful avoidance of putting technology first in the automation process, at the same time I would like you to recognize that technology is also the enabling factor to the solution to certain classes of problems. Knowing how graph databases work, for example, opens your mind to possible solutions to particular business problems that you would not have otherwise. An effective practitioner is not only a keen observer of the surrounding world through notes and diagrams but is also a master of the "art of the possible" through technology. Do not swing the pendulum so far the other way that you shun the work of staying on top of technological developments in the field.

A great practitioner has a spacious worldview. By that I mean they are not constrained by their company or industry in thinking that the microworld of their surroundings is so special and unique that you cannot learn from other companies and other industries altogether. I've seen many a company cloak itself in the idea that their company or industry is simply so sophisticated and distinct that its business is of a different Calculus from the rest. Solutions to common business problems outside—the management of risk, the optimization of supply chains, the intricate configuration of product portfolios, hold no lessons for them. This is dangerous thinking. As someone who has had a rare front-row seat looking at companies and industries side by side in time I can tell you that there is much to learn by becoming a true "student of business" through a traversal of industries to see what can be creatively imported from elsewhere. Always be a student of business everywhere you go.

You will know when you have become a full-fledged practitioner of automation. You will look at the world differently. You will read news articles with a different, clearer view. It is as if you put on extra powerful glasses to see things in your mind's eye that could not see before. It is like having... a *superpower*. You will notice right away how you simply think differently from those around you. You think in systems, and in pictures of systems. You find yourself constantly prodding colleagues to raise meeting conversations to the "big picture level" or discussing unintended consequences of a seemingly positive action. You will ask uncomfortable questions that no one else thinks to (or dares to) ask. You will see workers in action, performing their daily tasks, and in your brain you are already imagining the automated system that would operate in its place. It is precisely this quality of thinking that was my goal for you in reading this book.

Do you have a thick skin? I hope so, you will need it. Thinking in systems, wanting to experiment with ideas is not going to sit well with most people who live outside of the tail of the Power Law distribution. You will encounter unbelievable resistance to automation from people who should know better. Perhaps your biggest foe is the IT Department.

Ironically, Information Technology is the most change-resistant, bureaucratic corporate group anywhere. Rather than promote new uses of technology, they see themselves as guardians of some arbitrary "standard" architecture that is ideally suited for reinforcing the status quo and nothing else. In some organizations they are incredibly powerful in spite of their unjustifiable budgets and their equally poor performance. Your only defense against this corporate disease is the business itself—those functions inside the business side of the firm that are motivated to embrace automation for its benefits. A business sponsor armed with your carefully prepared diagrams, talking points, and ROI models can sometimes trump IT's need to squash anything that smells remotely like productivity.

To those CxOs reading this, I have this to say...

You don't have to be told that running any modern organization these days is hard work. You live it. It feels like battling a hydra—cut off one head and another swings around to attack you from a different angle. The complexity and chaos are unrelenting. Still, in a good year, the company magically hangs together and generates solid value. What you yearn for is *sustainable* growth, not the fits and starts of one-off initiatives scattered across functions. Automation is not a panacea, but it most definitely represents an enterprise-wide constellation of practices that deliver sustainable value to the firm if done correctly. My definition of "correctly" is laid out before you in these pages. To allow automation to flourish, however, you must clear the inevitable roadblocks to end-to-end experimentalism, and that often means championing exceptions to the company's norms—no easy feat, but essential.

The System Dynamics community centered around the Sloan Business School at MIT suggests that the role of senior leadership is to come up with an architecture or design for an organization that respects and even exploits the power of feedback loops. Their insights have led to some of the most profound transformations of businesses the world has seen, along with fundamental insights into how ideal organizations work. I second their idea, and in fact I am suggesting one can go even further by incorporating automation into those very feedback loops to make them even more effective, more responsive, and more durable. The clever CxO of today would be well served by a good dose of learning about Systems Thinking and System Dynamics [1].

For those of you attached to non-profit organizations I hope you took away from the book deeper lessons about automation and its efficacy that are perpendicular to the goal of making profits. Non-profit organizations deserve the same operational rigor that their for-profit counterparts enjoy, for it is the "doing more with less" imperative that drives non-profits to accomplish important contributions to our society. A non-profit organization that embraces automation will be a better non-profit organization, period. The level of investment to get there does not have to be prohibitively high. You have no excuse these days not to consider automation as a means to scale your mission across far more benefactors.

In my own professional career, the times where I had an explosion of creativity and productivity were the periods where I was blind and dumb. Blind to the limitations and constraints, dumb to all the plausible "go wrong" paths. I was ignorantly optimistic. In these days of social media and the Internet serving the world's raw views on your doorstep, it is incredibly easy to be overtaken by all of the negative views of … anything. If you are going to accomplish something meaningful in this world, you must be blind and dumb every once in a while. That is especially true of automation.

It has been an honor to be with you. Keep learning. Stay alert. Take time to think. Tell stories. Ask uncomfortable questions. Use your brand-new superpower to go out and make a difference for those around you.

Above all, best of luck.

Bibliography

1. John Sterman, *Business Dynamics* (McGraw-Hill Education, February 2000).

Index

© The Editor(s) (if applicable) and The Author(s) 2019
G. E. Danner, *The Executive's How-To Guide to Automation*,
https://doi.org/10.1007/978-3-319-99789-6